PIMLICO

349

# THE FLOATING EGG

Roger Osborne studied geology at Manchester University in the early 1970s. He then worked as a publisher of scientific, medical and technical books before becoming a full-time writer in 1992. He is the co-author of *The Atlas of Earth History* and *The Atlas of Evolution*, and lives in North Yorkshire.

# THE FLOATING EGG

## Episodes in the Making of Geology

ROGER OSBORNE

PIMLICO

Published by Pimlico 1999

6 8 10 9 7 5



First published in Great Britain by
Jonathan Cape Ltd 1998
Pimlico edition 1999

Pimlico
Random House, 20 Vauxhall Bridge Road,
London SW1V 2SA

Random House Australia (Pty) Limited
20 Alfred Street, Milsons Point, Sydney,
New South Wales 2061, Australia

Random House New Zealand Limited
18 Poland Road, Glenfield,
Auckland 10, New Zealand

Random House South Africa (Pty) Limited
Endulini, 5A Jubilee Road, Parktown 2193, South Africa

Random House UK Limited Reg. No. 954009

A CIP catalogue record for this book
is available from the British Library

ISBN 0-7126-6686-9

Papers used by Random House UK Limited are natural,
recyclable products made from wood grown in sustainable forests.
The manufacturing processes conform to the environmental
regulations of the country of origin

Designed by Peter Ward

Printed and bound in Great Britain by
Mackays of Chatham PLC

For Jan

# Contents

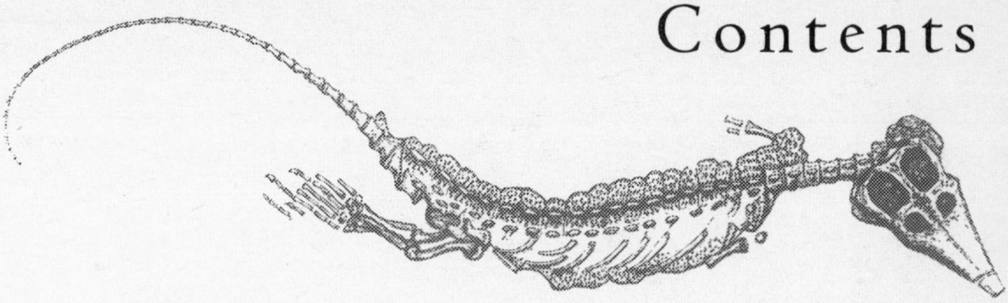

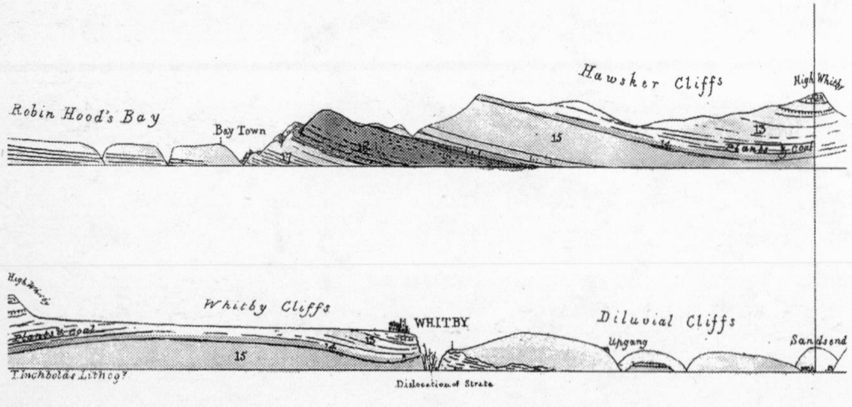
Hawsker Cliffs
High Whitby
Robin Hood's Bay
Bay Town
15
13
Whitby Cliffs
Plants & Coal
WHITBY
Diluvial Cliffs
Upgang
Sandsend
T. Inchbold's Lithog.
Dislocation of Strata

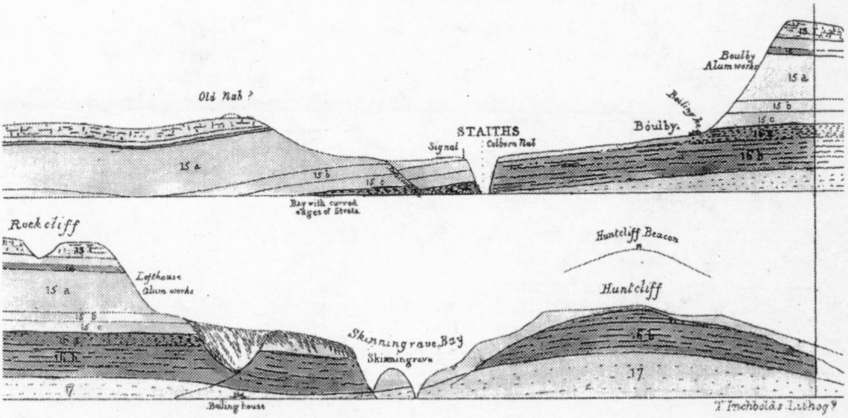
Old Nab ?
STAITHS
Signal
Colborn Nab
Boulby
Alum works
Boulby.
15 a
15 b
15 c
16 b
Bay with curved edges of Strata
Rock cliff
Lofthouse
Alum works
Huntcliff Beacon
Huntcliff
Skinningrave Bay
Skinningrave
17
Boiling house
T. Inchbolds Lithog.

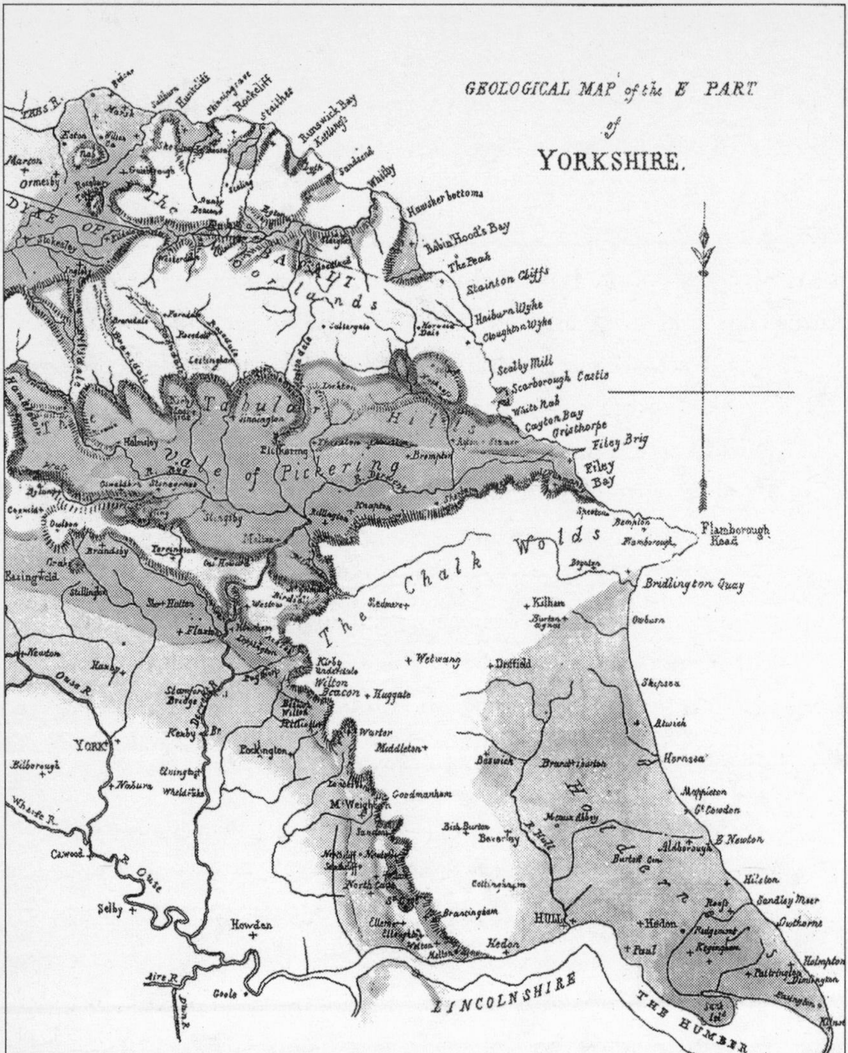


*The setting for the episodes related in this book (taken from* Illustrations of the Geology of Yorkshire *by John Phillips, 1829).*

# Preface

In the next few hundred pages all is not quite as it seems. The pieces of writing that follow are all concerned with the origin and development of geological sciences in a particular part of the world (see the map opposite). Each piece is a separate story, and the reason for their co-existence will become apparent as they are read. The stories are arranged in approximate chronological order of the events they describe, though the 'reptile tales' follow a separate order from the other pieces. Having made that rule for the organisation of the book, I then broke it several times in order to make the whole thing somehow more agreeable. Let us just say that the main pieces begin with the alum trade in the early seventeenth century and end with the discovery of ice-age lakes in the early twentieth. There is also a brief chronology of events at the back of the book.

Some confusion might arise over the authorship of these stories, and the reality or otherwise of the events they describe. So let me begin by asserting that everything in this book is true. All of the events described took place, most of them 'exactly' as documented. Every document that is quoted from is real, and is listed in the Notes and Sources. In those pieces where the author is a real 'other' person, his or her name and full reference is given. Those pieces which are written by me as a factual historical narrative will be obvious to the reader. In only three longer pieces is the narrative or the narrator in any way fictional, as follows.

In 'A Journey to a Birth' the narrator is Samborn Palmer. He is a real historical figure who accompanied William Smith on the journey which the story relates. The story is 'fictional' but the events are as described by

William Smith to his nephew John Phillips, who then wrote them down (Smith was a practical man who was averse to writing). 'The Strange Case of the Hyenas' Bones' is narrated by a fictional character of my own invention, John Foster. Apart from Foster and Dawson, the other characters and the principal events in this story are as described in William Buckland's own works, and by subsequent historians of geology. In 'From under the ice' the narrator is fictional, as is the story, and really needs no further explanation.

All of the stories in this book draw heavily on the work of other authors, some to the point where they are simple reproductions of what others have written. Because of this intimate relationship between the content of the stories and the primary source material, I have included a full chapter-by-chapter list of sources, and in some cases further explanations of individual chapters, at the end of the book. There are, though, some individuals whose help has been so important in the conception and writing of this book that they should be listed here. I therefore unreservedly thank Mike Benton of the University of Bristol for his initial encouragement and for the information he provided; Christopher Toland for sharing his enthusiasm for, and knowledge of, the history of geology in Yorkshire; and most of all my editor Will Sulkin, whose nurturing of this book and its author has been above and beyond the line of duty. In addition Messrs R. L. P. and J. G. Pickles of the Whitby Literary and Philosophical Society, together with the staff of their museum and library have been of great help, as have the staff of the libraries of the Geological Society and the Royal Society. They have my sincere thanks.

Roger Osborne
Scarborough, 1998

chapter one

# Dangerous Rocks

## *'The most singular accident'*

"The coast from Saltburn to Scarborough, and even to Bridlington Quay, is generally high and bold, except in the bays and inlets. At Huntcliff, Rockcliff, Kettleness, Peak, and a few other places, the cliffs are lofty, and in some parts precipitous. Hence these shores are not only dangerous to mariners in stormy weather, but cause many fatal accidents to others who frequent them. The most singular accident that ever happened on the coast, occurred about 15 years ago, under the high cliffs a little to the west of Staiths. While two girls of the name Grundy, belonging to Staiths, were sitting on the *scar*, or rocky beach, with their backs to the cliff, a splinter, which by striking against a ledge had acquired a rotatory motion, fell from the cliff, and hitting one of the girls on the hinder part of the neck, severed her head from her body in a moment, and the head rolled to a considerable distance along the scar."

George Young, 1817

chapter two

# But Dreames

## *In search of the alum-maker's secret*

"The World hath beene much abused by the opinion of *Making of Gold*: The Worke it selfe I judge to be possible; but the Meanes (hitherto propounded) to effect it, are, in the Practice, full of Errour and Imposture; and in the Theory, full of unfound Imaginations. For to say, that *Nature* hath an Intention to make all Metals *Gold*. And that, if she were delivered from impediments, shee would performe her owne Worke; And that, if the Crudities, Impurities, and Leprosities of *Metalls* were cured, they would become *Gold*; And that a little *Quantity* of the *Medicine*, in the Worke of *Projection*, will turne a Sea of the *Baser Metall* into *Gold*, by *Multiplying*; All these are but dreames."

Francis Bacon, 1626

The etymology of 'alchemy' is instructive and deceptive. The origins of the word lead us into the market-places and academies and the secret places of the eastern Mediterranean, where Greek apothecaries learned from Egyptian travellers, and Arab physics made practice from the natural philosophies of scholars. Alchemy is defined by us, in a striking piece of historicism, as 'the infant stage of chemistry'.* All are agreed that its origin is the Arabic word *al-kimya.* But the origin of *kimya* is disputed,

* *Chambers 20th Century Dictionary*, 1983 edition

though it is probably Greek. The Greek language has several similar words. *Khemia,* which means 'black land', was the name given to Egypt by the Greeks, according to Plutarch. They were following the Egyptians who called their own land *Kemi,* the black land. Khemia is then possibly 'the Egyptian art' or it may mean the 'black art' or 'dark practices'. Another Greek word *khymeia* means fusion, perhaps the combining of precious metals. Then there is *Chymes,* who has been suggested as the inventor of alchemy. And finally the word *chyma* which means fluid, or the art of pouring.

But the principles of alchemy – the freeing both of substances and humans from their corrupt temporal existence, into the eternal purity of gold and immortality – are older than the word. Records of these in India and China are as ancient as written language.

*Cliffs east of Whitby. The cliffs of the Yorkshire coast were transformed from base rock into apparently limitless riches.*

# FRAGMENTS TOWARDS A HISTORY OF ALUM

"Not less important, or indeed dissimilar, are the uses made of *alumen*; by which name is understood a sort of salty earth . . . the white alumen being employed in a liquid state for dyeing wool bright colours, and the dark-coloured alumen, on the other hand, for giving wool a sombre tint. Gold is purified with black alumen. Every kind of alumen is from a limus water which exudes from the earth. The collection of it commences in winter, and it is dried by the summer sun. That portion which first matures is the whitest. It is obtained in Spain, Egypt, Armenia, Macedonia, Pontus, Africa and the islands of Sardinia, Melos, Lepari, and Strongyle; the most esteemed, however, is that of Egypt, the next best from Melos."

Pliny (23–79 AD)

"Alum (*fan shih*)

It comes now from Hsi-chhuan [Szechuan] in the north of I-chou, crossing the (Yellow) River from the west. Its colour is greenish-white. The crude stuff is called 'horse-tooth alum', but after purifying it becomes very white. People in Szechuan often mistake it for nitre (*hsiao shih*), but it is white alum . . . The Manuals of the Immortals say that it can be taken by itself, and it is used also in elixir formulae. After being ground in water it can be combined with plant drugs, then boiled and heated to dryness; it is good for toothache but spoils the teeth if used in excess. This shows that it is injurious for bones, so I doubt the statement that it hardens bones and teeth."

Thao Hung-Ching (456–536 AD)

"Methods for preparing the elixir of life, which are known to be wrong or dangerous, but which are popular among the people.

– Boiling the ash obtained from burning mulberry wood.
– Mixing common salt, ammonium chloride and urine and evaporating to dryness.
– Digesting saltpetre and quartz for a long time in a gourd and using the product.
– Boiling saltpetre and blue-green rock salt in water.
– Making an egg-shaped container of silver to hold cinnabar, mercury and alum.
– Using iron rust and copper as ingredients for an elixir called 'golden flower'.
– Heating mercury together with malachite and azurite.
– Heating realgar and orpient.
– Heating black lead with silver.
– Burning together dried dung and wax.

Some alchemists have heated sulphur together with realgar, saltpetre and honey, with the result that their hands and faces have been scorched, and even their houses burnt down."

*Chen Yuan Miao Tao Yao Lueh (Classified Essential of the Mysterious Tao of the True Origin (of Things))* (8th or 9th century)

"The Nature of Alum

Alum is an oil which has been coagulated by the dryness of the earth, and there are many mines of it. The son of Gigil of Cordoba said that Gigil had a mine to the north of Cordoba in a place which is called Neeris; in the neighbourhood of old Cordoba.

And there is in its nature heat and dampness; and it is called lime and *sicum*. And indeed it retains every volatile spirit and it cleanses metals with a good cleaning and embellishes them, and increases their dyeing. Nevertheless it dyes a white thing black as long as it is not made active from the nature of the white thing, and it is not destroyed; and it makes a white thing red as long as it dyes them with a good natural redness, and it remains dyed for ever . . .

"The Method of Jamen Alum

The method of using it is, that you take some pure white woolly Jamen alum, and pound it very well, and put it in a glass jar; and you pour over it four times as much boys' urine. After boiling them down and depending on the age of the urine, boil it over a small flame with the alum, and stir it well with wood, and set it aside until the sediment of it settles down on the bottom. Then filter it gently and put aside the urine with your collection of secrets."

Rhazes (Abu Bakr Muhammad al-Razi), (10th century)

"A hundred years of living is as a spark from a stone, and the course of life a bubble floating on water. Those who think only of profits and emoluments, seeking worldly prosperity and glory, will soon find their faces turning pale and their bodies withering. Even if they have piled up riches mountain-high, may I be allowed to ask whether they can buy off the Messenger of Death . . .

Compound not the three yellow substances (sulphur, orpiment and realgar), neither the four magical things, (alum, cinnabar, lead and mercury); and make no search for particular plants (as elixir ingredients) for they are still further from the genuine (medicine)."

Chang Po-Tuan, (d. 1082), *Wu Chen Phien* ('Poetical Essay on Realising the Necessity of Regenerating the Primary Vitalities')

"[alum (*fan shih*)] can turn iron into copper . . . The yellow and dark yellow sorts of alum are called 'bird-droppings alum'; these are not used in pharmacy but are only suitable for the plating (*tu*) (of metals). When they are added to processed copper (*shu thung*) (powder), being made into a paste with crude vinegar and smeared on the surface of iron, the iron is all turned to the colour of copper. But although the outside colour becomes coppery the material inside remains quite unchanged."

*Ching-Shih Cheng Lei Pei-Chi Pen Tshoa* (the *Classified and Consolidated Armamentarium of Pharmaceutical Natural History*) (1083)

## MINERALS AND SECRETS

How would we know the alum-maker's secret? This, at least, we can answer – it is now written down, freely available, much wondered at; and entirely useless. Overtaken by history, it has served its time, and in its time it possessed the power to turn the basest rock into a substance worth more in the world than gold itself.

The source of alum is almost everywhere, in almost every rock. It is the knowledge of its manufacture that is wanted. We meet alum frequently enough in the pages of history – a magical element in Chinese and Arabic alchemy, a reagent in early chemistry, a mordant and tanning agent, a medicine, an essential part of the agricultural manufactories of early modern Europe – a chemical that every flexing, strutting state needed to give substance to its virility.

The name of this substance means little to us now. Our society is disaggregated enough for us to live well without knowing much of what enables us to do so. Alum is still a vital chemical in every developing and industrial society. But the problem of obtaining alum has been solved, and so the substance has disappeared from view. Invaluable yet invisible, it is the concern of just a few industrial chemists.

To our predecessors the opposite was true. Alum was essential, and scarce. It was therefore a strategic commodity, and a potential source of vast wealth and political influence. Anyone who knew how to make alum, who found how to transform stone into wealth, who had the alum-maker's secret, might hold half the world to ransom.

But can history – by which we mean the written record of the past – bring us face to face with the origins of something which was never, being secret, transcribed while it was of value; something that could only emerge into view once chemistry caught up with alchemy, and the keeping of secrets was replaced by the sharing of knowledge; when science began to understand technology?

## THE 'HOLY FATHER'S MARCHANDISE'

At some time in the early decades of the fifteenth century John de Castro, an Italian whose origins are unknown, set up a cloth dyeing factory in Constantinople. Venice, Genoa, the Papacy and other city states of Italy traded in manufactured goods and raw materials, including alum, with the cities of the region we now call Asia Minor. Italians moved east and set up as traders, shippers and manufacturers. John de Castro used alum as a dye fixative in his work, he knew its value, and he learned something of how it was made. It is thought that Italians were involved in alum manufacture near the port of Smyrna (now called Izmir), but their names are unknown.

The area around Constantinople was the remnant of the Byzantine lands, while the cities of the eastern Mediterranean and Persia were ruled by families whose influence spread only a little further than their own city walls. Vast areas of countryside were controlled by nomadic tribes. In a process endlessly repeated in the history of human affairs, a unifier emerged to persuade the nomads of the outer lands of their sufferings, and to lead them against the cities of the centre. The leader was called 'Uthman, which in Turkish is Osman, and this is represented in English as Ottoman.

The power of the Ottomans originated in Anatolia and spread west into south-eastern Europe, and eventually east and south into Syria, Egypt and western Arabia. By the middle of the fifteenth century the city of Constantinople with its vigorous mix of Islamic, Christian, Arabic, Byzantine and European peoples, was isolated. In 1453, one of the most symbolic dates in the intertwined histories of Europe and Asia, an event both unimaginable and inevitable happened – Constantinople fell to the Ottoman armies of Mehemmed II. The illusion of the Holy Roman Empire in the East was, at last, dissolved. John de Castro, together with most other Europeans, left Constantinople. He was able to return to Italy and gain employment under the Apostolic Chamber, part of the administration of the Papacy – spiritual ruler of Christendom and secular ruler of central Italy.

The Ottoman Empire now controlled Europe's alum supply. The textile manufactories of Italy needed alum, so they bought it from the Turks. But they hated their dependence on this troublesome, expansive, unchristian power, and desperately sought their own supply. A few small works were started, more in hope than knowledge, and these foundered.

John de Castro began his own investigations in the area around Rome. The Tolfa hills, north-west of the Holy City, had been a source of minerals from the times of the Etruscans, twenty centuries earlier. De Castro found rock at Tolfa which looked the same as those he had seen being worked for alum near Constantinople and Smyrna. He made tests, which convinced him that he was right. 'Experts' were sent from Rome, who, it is said, 'shed tears of joy, kneeling down three times, worshipped God and praised His kindness in conferring such a gift on their age.' The works at Tolfa, not sixty miles from the Vatican itself, were established as a Papal monopoly, and were producing alum by the year 1459 – a mere six years after the fall of Constantinople. It was, in the view of Christian Europe, a victory for God over the infidel.

There was enough alum in the Tolfa hills to supply the whole of Europe for a hundred years. Alum might have carried on bringing gold to the Pope's coffers for hundreds of years to come, had not the Church encountered its own difficulties. Europe's only source of alum belonged to God's sole representative on Earth. As one monopoly unravelled, so did the other.

In 1517 it was decided that the Church of St Peter in Rome was in need of rebuilding and repair. Notices were sent out to the cardinals and archbishops of all the countries of Christendom, that money should be raised from their flocks, so that the parish church of the Catholic faith should be made glorious again. So the Archbishop of Mainz, a historic city on the banks of the Rhine, dispatched a Dominican friar by the name of Tetzel to the town of Wittenberg. There he was to sell indulgences, so that the faithful might buy remission from the punishment of their sins, and help the Church on Earth at the same time. We know only enough of Tetzel to say that he took his proper place in a story that was necessarily beyond his comprehension.

It was another story of the type that appeals to our misguided sense of the potential of every man to shake the world. A figure of impregnable authority, carrying the force of the spiritual and secular world in his person, approaches the moral and geographic milieu of a slightly troublesome yet impotent irritant – and is brushed aside by . . . what? The requirements of history? By the needs of a widely diffused force of dissent to find a single point of expression? Or by the moral authority of one unexceptional, exceptional man: Martin Luther.

Six years before Tetzel's arrival in the town where he taught and preached, Luther had made his pilgrimage to Rome, where he had been overwhelmed and sickened by the luxury, wealth and power games that were the daily life of the Papal court. He returned to the Germanic states and preached a doctrine of grace and faith based on his readings of St Paul and St Augustine. In particular he spoke against the sale of indulgences.

The arrival of Tetzel forced Martin Luther to carry his arguments further. As every European schoolchild is now taught, he drew up a list of forty-nine arguments, or *theses,* against the sale of indulgences and nailed it to the door of Wittenberg church. Within forty years Protestantism in one form or another was established in the major states of northern Europe, under the doctrine of Augsburg – *cujus regio, ejus religio,* the religion of the ruler is the religion of the state.

The dependence of the cloth and leather industries of the whole continent on one supplier of alum was unstable before the Reformation. Afterwards it was impossible. In 1536 a German physician named Georg Bauer, who, in the fashion of the times gave himself a 'Latinised' name, Georgius Agricola, published a book entitled *De Re Metallica.* Further editions of the book appeared in 1561, 1621 and 1657. It was the first attempt to produce a systematic guide to the production and properties of different metals and minerals, and included details of three different methods of producing alum. Though Agricola was able to outline the process, he did not mention the alum-maker's secret, without which the process is unworkable. He may have considered it unimportant, or he may not have known of its existence. He might have known about it, but been unable to discover what it was. The secret remained.

The bringing of alum-working to the cliffs of Yorkshire, which was the first and only source of alum in Britain from 1600 until 1870, is documented, but not detailed. So relatively late in history our vision is clouded by a perplexing mix of romance, prejudice, myth-making, and the aggrandisement of important men. It seems that the growth of printed matter can be a false guide to the paths of history – not all writers felt the need to record what was true. To their unreliable bequests we can add our own 'knowledge' of documentation and of scientific laws – these, at least, have little variance with time.

By the late sixteenth century there were alum works in Germany and Flanders, but none in the British Isles. England, its commerce expanding, its alliances fractious, was desperate for a dependable source of its own. (Later reports that Henry VIII married Anne of Cleves to get access to the Flanders alum are an exaggeration, not an impossibility.) A licence granted in 1560 to a W. Kendall to manufacture alum from rocks in Devon and Cornwall is a sign of activity and of ignorance. There is no rock in these counties from which alum could practicably be made, as a rudimentary knowledge of Agricola's work would have shown.

The first document that betrays the anxieties of the English court is a Letter Patent granted by Elizabeth I in 1565.

> "Elizabeth, by the Grace of God, Queen of England, France, and Ireland, Defender of the Faith &c. To all Men to whom these Letters Patent shall come, Greeting. Where heretofore we have granted Privileges to Cornelius de Voz, for the Mining and Digging in our Realm of England, for Allom and Copperas, and for divers Ewers of Metals that were to be found in digging for the said Allom and Copperas, incidentally and consequently without fraud or guile, as by the same our Privilege may appear."

This 'licence' stimulated some attempts – Alum Bay on the Isle of Wight is named for a fruitless sixteenth-century quest – but prospectors clearly had no idea what they were looking for.

We next come upon alum in John Aubrey's *Lives of Eminent Men*, written in the seventeenth century, but unpublished until the nineteenth. Aubrey writes well of a North Country gentleman, Sir Thomas Chaloner, who was known enough to be appointed tutor to Prince Henry, eldest son of King James. Aubrey approves of his subject.

> "He was a well-bred gentleman, and of very good naturall parts, and of an agreable humor. He had the accomplishments of studies at home, and travells in France, Italie, and Germanie . . . He was as far from a puritan as the East from the West."

So far, so good. But now Aubrey ventures into the territory of the technical. Chaloner was, he writes,

> " riding a hunting in Yorkeshire (where the allum workes now are), on a common, he tooke notice of the soyle and herbage and tasted the water, and found it to be like that where he had seen the allum-workes in Germanie. Whereupon he gott a patent from the King (Charles I) for an allum worke (which was the first that ever was in England), which was worth to him two thousand pounds a year or better: but tempore Caroli I some courtiers did thinke the profitt too much for him, and prevailed so with the king, that, notwithstanding the patent aforesayd, he graunted a moetie, or more to another (a courtier), which was the reason that made Mr Chaloner so interest himselfe for the Parliament-cause, and, in revenge, to be one of the king's judges."

Allowing for technical inaccuracies, we need look no further than this passage for the intertwining of commerce, politics and the riches of the earth – and of the penalties for misjudgements in any of these. The patent was, in fact, granted to Chaloner by James I (or even Elizabeth), who used this method for raising money without the inconvenience of assembling Parliament – a habit that was to cost James's son his kingdom and his life. Maybe Aubrey is seeking a simple reason for a complex matter – why should an English gentleman landowner turn into the ardent republican that Chaloner became, even to the point of sitting in fatal judgment on his own king? But if King Charles treated Chaloner unjustly, or gave him cause to think so, and if he behaved similarly to others, he paid the ultimate price for his divine exercise of power.

Chaloner, an energetic, ambitious scholar-merchant, is our bridge between the old southern, Catholic-dominated, Romish Europe with its debts to scholasticism and classical Rome; and the new, northern, Protestant or even atheistic world, owing intellectual allegiance to Platonic Greece, and to its own self-confident rationalism. But Chaloner has enough of the old world to be cloaked in a swirl of romance. It was Chaloner, it seems, who brought the alum-maker's secret to England.

The managers of the Papal alum works at Tolfa were well aware of the value of the secret they held – the secret of turning common rock into precious alum. Why else were the workmen kept in virtual captivity? To protect the great secret was to protect the wealth of the Papacy.

On a tour of Europe, during which he saw mineral workings in Germany and Italy, Sir Thomas Chaloner sailed into the harbour of Civitavecchia, the alum port of the Papal states. It is said that he visited the alum works, though whether he was permitted to see everything at close hand is doubtful. Whatever he did see persuaded him that he needed to know what was going on inside the alum houses. He therefore did the simplest, and most dangerous thing – he persuaded one, or perhaps two, of the Italian workmen to abscond with him. The legend says that he smuggled them on to his ship, anchored at Civitavecchia, in a barrel, and sailed for England before they were missed. George Young, writing in 1817, says the men were called Russel, which must have been an Anglicisation of their Italian names.

Now Chaloner had the alum-workers, but did he have the alum-worker's secret? It is said that, for his crime, Chaloner was excommunicated. Though there is no specific evidence of this, we can judge the likelihood by this passage from Bishop Latimer's sermons (1596); alum was, we need not doubt, a precious and holy matter:

> "I heard a great while ago a tale of one. He hath travelled in mo countries than one. He tolde that there was once a Pretor in Rome, Lord maior in Rome, a rich man, one of the richest marchants in all the Citie, and suddenlie he was cast into the castell Angell. It was heard of, and everie man whispered in anothers eare. What hath he done? hath he killed any man? No. Hath he medled with Alam, our holy fathers marchandise? No. Hath he counterfeited our holy fathers Buls? No. For these were hie treasons. One rounded another in the eare and said: *Erat dives*, he was a rich man. A great fault."

## ENGINEERING ALCHEMY

Chaloner must have known, or been told, that his shales had potential for the manufacture of alum. Agricola's *De Re Metallica* was available to diligent scholars, and its contents would have reached England by the late sixteenth century.* Even with this knowledge and the expertise of his Italian workmen, it took several years for the works at Belman Bank to produce pure alum. Once they did, the value of the alum shales was an open secret.

* We have no direct evidence that Chaloner had read Agricola's work, and should be cautious about assuming a complete diffusion of technical knowledge through published books.

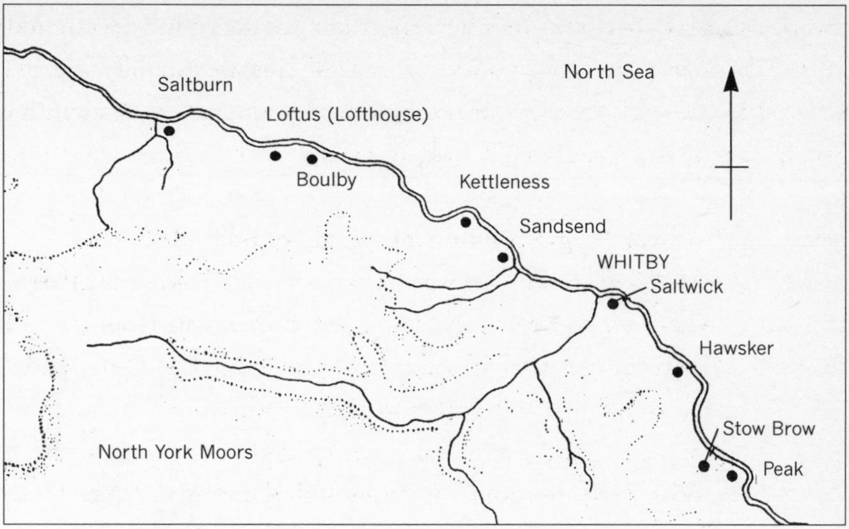


*Sites of alum works on the Yorkshire coast over nearly three centuries (after Marshall, 1995)*

The shales occur in a band up to 35 metres thick in the Lower Jurassic, Toarcian section of north-east Yorkshire. They outcrop inland in a few places, as at Belman Bank. But their greatest exposure is on the series of towering cliffs that run from Loftus in the north-west, through Boulby, Port Mulgrave, Kettleness and Sandsend to Whitby; and then on through Saltwick, Hawsker and Robin Hood's Bay to Blea Wyke in the south-east. Every landowner whose estate encompassed a portion of that desolate, beautiful, useless coastline, was instantly an industrial baron. Chaloner and his heirs had no monopoly on the outcrops of alum shale, nor on the alum-maker's secret. It was impossible to contain knowledge at such close quarters. The secret spread along the coast, and so did the alum works. In the two and a half centuries from 1620 to 1870, the cliffs were ripped open and thrown on to the beach. Millions and millions of tons of rock were taken out by pickaxe and wheelbarrow. By the early part of the nineteenth century production was over 3,000 tons of alum per year. Twelve tons of shale produce one ton of alum. At least three tons of overburden would have been removed for every one ton of shale. The pretty seaside town of Sandsend was built solely for alum workers. And even the famous port of Whitby owes more than it will admit to this first industrial-scale chemical technology. Historians will glibly say that Whitby is an ancient fishing port, prospering and expanding with its fisher fleet and through shipbuilding. But until the alum trade came, hungry for coal and wanting a harbour to ship out its precious mineral, Whitby was a fishing village with a ruined abbey incongruously attached, not a port.

Alum is a double sulphate – aluminium and either potassium or ammonia form the positive ion, sulphate the negative. The two most common elements in the Earth's crust are aluminium and silicon; almost all rock-forming minerals (olivines, amphiboles, pyroxenes) are aluminosilicates. These are extremely stable structures. The task of the alum-worker is to break down the common aluminosilicates in the presence of sulphur, in a way that allows the aluminium to be freed to form a sulphate. In order to enable this to happen an alum source must have three crucial ingredients, apart from the aluminosilicates that occur in every rock; it must contain sulphur (usually as iron pyrite), water and carbon. The alum shales of the Yorkshire coast have these ingredients in exactly the right proportion.

The initial process of alum-making is physically crude, but chemically quite complex. The shales were hacked from high up on the cliffs with pickaxes, and tipped off wooden platforms on to layers of brushwood on the beach. When the piles of shale were 50 feet high, the brushwood was lit. The carbon in the shales kept the pile burning, usually for about nine months.* Piles often grew to be as big as 100 feet high and 150 feet long.

* Kendall, writing in 1924, says that the main shale band at Boulby was still burning, even though the works were abandoned in 1871 – more than fifty years previously.

*The village of Sandsend near Whitby was built for the alum trade. The disused alum houses are in the village, while the cliffs in the background have been torn down by the alum quarriers. Since this photograph was taken, the railway viaduct has gone and Sandsend is now a pretty seaside village.*

The burning, or calcining as it is properly known, sets a chain of reactions in train. The iron pyrite ($FeS_2$) in the shales loses half its sulphur as sulphur dioxide ($SO_2$); the remainder becomes a black iron sulphide ($FeS$); this sulphide rapidly absorbs water and oxygen to form ferrous sulphate ($FeSO_47H_2O$); some of this ferrous sulphate remains and some becomes oxidised in the high temperature and gives out sulphuric acid ($H_2SO_4$); this sulphuric acid then reacts with the aluminosilicates (the bulk of the shales) to produce aluminium sulphate ($AlSO_4$).

When the piles were broken open after nine months of calcining, the shales had been converted into a powdery mixture of ferrous sulphate and aluminium sulphate, with some insoluble silicates enduring as slag. Clean water was poured through the piles, dissolving the sulphates and leaving the silicates behind. This precious solution of sulphates was then channelled to the alum houses. Now that the alum-workers had a solution of aluminium sulphate, all that remained was to convert the aluminium sulphate into a double potassium-aluminium sulphate or ammonium-aluminium sulphate, which is alum. This was easy enough. They simply added potash in the form of kelp or potassium chloride, or ammonia in the form of human urine. Urine was brought in specially adapted ships from Sunderland and Newcastle. The same ships then returned carrying alum.

But one overriding problem remained for the alum-workers. Once the potash or ammonia was added to the solution, the alum was formed. But the alum was in solution with ferrous sulphate – the intermediate product of the calcining process. If the mixture was simply allowed to crystallise out of the solution, the ferrous sulphate impurity would render it useless. All of the quarrying, all the calcining, the transportation of the potash, the urine and the precious liquor was of no use whatsoever, unless a way could be found of separating the alum from the ferrous sulphate. But if a way could be found, then the cliffs of Yorkshire would be an inexhaustible supply of wealth, a bottomless crock of gold. The method of separation is the alum-maker's secret.

Add a soluble salt to water and it will dissolve. Keep adding more and more salt – you will reach a point where the water won't dissolve any more and the surplus salt will remain as powder or crystals. The solution is then saturated. But different chemicals are soluble to different degrees, which means that they reach their saturation points at different concentrations. The alum-workers were faced with this situation, but in reverse. They had their solution of sulphates, and they wanted the salts which it contained to crystallise out. This is simple enough: you increase the concentrations of the salts by removing water from the solution – by boiling it away. The liquor, or 'mothers' as it was known, was channelled into great open pans in the alum houses, which were heated from beneath by coal fires. Fortunately the saturation point of alum (the concentration at which it would begin to crystallise out of the evaporating solution) is a lower concentration than the saturation point of ferrous sulphate. Alum is simply not as soluble in water as ferrous sulphate. So the alum-workers boiled the 'mothers' until they reached the concentration at which alum was over-saturated, but ferrous sulphate was under-saturated. The alum would then, almost miraculously, crystallise out of the solution, leaving the unwanted ferrous sulphate still dissolved.

Simple, except for one very big problem. Almost all chemicals, including alum and ferrous sulphate, dissolve to a higher concentration in hot water than cold. It was necessary for the alum-workers to heat the 'mothers' in order to evaporate the water, and achieve what seemed like the correct concentration. But if they simply heated the liquid until they could actually see the alum crystallising out, and then left the liquid to cool (which was what they wanted to do, since they could not keep the liquor hot while they waited for crystallisation) then the concentration of ferrous sulphate would have been too high in the hot liquid, and it too would crystallise out as the liquor cooled and it became less soluble. The only way to get the alum to crystallise out of the cooled liquor without the ferrous sulphate would have been to cool the liquid down to see if the process was working, and then if it wasn't, to heat it up again and evaporate more water. But if you heated it too much you got ferrous sulphate crystals forming along with the alum, which would render the product worthless. The whole process was redundant unless there was a reliable way to repeatedly estimate the concentrations of the salts in the hot liquor. When water, or any liquid, dissolves something, the density of the liquid is changed. The easiest way to measure the amount, or concentration, of the dissolved salt is, by some means or other, to measure the specific gravity (i.e. the density of a liquid relative to pure water) of the liquor. Through centuries of whispered conversation, or years of trial and error, or one moment of inspiration, someone figured out a way of knowing when the specific gravity of the hot alum liquor was exactly right. Someone, somewhere, some time came to the realisation that if a hen's egg was placed in the liquor, it would float to the surface at the precise moment when the concentration of the solution reached the optimum level. The alum-maker's secret was a floating egg.

It seems we will never meet, in the pages of documented history, the person who first stumbled on the method. Was it a Chinese alchemist from east of the Yellow River; or an Arabic philosopher from Khurusan or Córdoba; or a Turkish merchant from Smyrna; or an Italian cloth dyer from Cerveteri or Tolfa? Even if the tales are embellished in the telling, we can be sure that it would have been worth Thomas Chaloner crossing half the world to find out how to do it – and worth a Pope's fortune to stop him. Or perhaps the discoverer of the floating egg was a Yorkshireman, determined to improve the productivity of his works, using an artisan's trick to usher in a modern industry.

Civilisations are built on what we hack and rip from the Earth, and in the ways we transform it. Alum was part of human history before it came to Yorkshire, but the scale of manufacture here enabled England to become a powerful agrarian-based economy centuries before the industrial revolution. Alum-making transformed common shale into uncommon wealth, making fortunes for kings and for landowners. It built towns, villages and great houses. It brought the sins of greed and despoliation to a remote corner of England. It may even have cost a king his head. And as suddenly as it came, the alum trade went. New methods of producing sulphuric acid, developed in the 1870s, meant that alum could be made more cheaply by a simple chemical process. The story of alum-making on the cliffs of Yorkshire was brought to an end by the beginning of another story.

The ghost of the alum trade is still there. The ruins of the alum houses, the stone liquor channels, the disused quarries, are dotted along thirty miles of coast. And the digging of the alum uncovered another, even stranger and more improbable story. This time the hidden treasures were not raw materials or precious stones or metals. They were the inhabitants of former worlds. The alum works uncovered fossilised skeletons of hundreds of marine reptiles at the time when the world was just beginning to be ready to receive and understand them. This coincidence helped us to think the unthinkable, and to re-imagine a past of previously unbelievable remoteness.

chapter three

# The First Reptile

## *The allegator on the beach*

In 1758 a strange event took place at Whitby. Two friends, Captain William Chapman and Mr Wooler (whose forename is lost to history) discovered some fossil bones embedded in the rock of the beach at Saltwick. They removed the bones and were able to identify them as the fossilised remains of the skeleton of an alligator. Remarkable though this was, the peculiarity of the event did not lie in the discovery itself – fossil bones must have been found on this coast for as long as the area was inhabited. But what happened next signified a change in the relations of the human mind to the natural world. Whereas previously the two men might have kept some bones as souvenirs and perhaps showed them to a few friends, now they did something else. They each wrote a formal report of their find – Chapman in the form of a letter to a friend – with the intention (we may safely assume) that these should be published in the *Philosophical Transactions* of the Royal Society. This monthly journal provided a forum for the exchange of information and ideas on all aspects of natural history. It was the original scientific journal and without it science would simply never have been able to develop in the way that it has. Chapman and Wooler wrote to the Royal Society because of the existence of the *Philosophical Transactions*. In doing so, they transformed themselves from beachcombers to natural philosophers.

*"An account of the fossile bones of an allegator, found on the sea-shore near Whitby in Yorkshire.* by William Chapman, communicated by Mr J. Fothergill.

> A few days since we discovered on the sea-shore, about half a mile from this place, part of the bones of an animal . . . The ground they laid in is what we call allum-rock; a kind of black slate, that may be taken up in flakes, and is continually wearing away by the surf of the sea, and the washing of stones, sand &c. over it every tide.
>
> The bones were covered five or six feet with the water every full sea, and were about nine or ten yards from the cliff, which is nearly perpendicular, and about sixty yards high, and is continualy wearing away, by the washing of the sea against it; and, if I may judge by what has happened in my own memory, it must have extended beyond these bones less than a century ago. There are several regular strata or layers of stone, of some yards thickness, that run along the cliff, nearly parallel to the horizon and to one another. I mention this to obviate an objection, that this animal may have been upon the surface, and in a series of years may have sunk down to where it lay; which will now appear impossible, at least when the stones, &c. have had their present consistence . . .
>
> The place where these bones lay was frequently covered with sea-sand, to the depth of two feet, and seldom quite bare; which was the occasion of their being rarely seen: but being informed that they had been discovered by some people two or three years ago, we had one of them with us on the spot, who told us,

that when he first saw it, it was intire, and had two short legs on that part of the *vertebrae* wanting towards the head. Altho' we could not suspect the veracity of this person, we thought he was mistaken; for we had hitherto taken it for a fish. But when we took it up, and found the *os femoris* [femur bone] . . . we had cause to believe his relation true, and to rank this animal among the lizard kind: by the length (something more than ten feet) it seems to have been an allegator; but I shall be glad to have thy opinion about it.

I am thy friend,
William Chapman *Sen*."

Wooler's account, published six months after Chapman's, contains more speculation on the origins of fossils. He follows the knowledge of the time, in stating that the rocks that were once lying above the fossil alligator, but had since been eroded by the sea, were formed at the time of the Universal Deluge – Noah's flood. The logic here was simple and powerful. The Bible was a description of the history of the world since, and including, its creation. Most of the Earth was made at the Creation, as told in the first chapters of Genesis. But animal life was created after the sea and the land, so how could fossil animals be buried beneath the land? The answer must lie in a subsequent episode in the history of the world. The Deluge is the only event described in the Bible with sufficient worldwide impact to have produced this effect. All 'secondary' – i.e. sedimentary or fossil-bearing – rocks were therefore believed to have been formed by the Deluge.

"*A Description of the fossil Skeleton of an Animal found in the Alum Rock near Whitby*. By Mr Wooler. Communicated by Charles Morton, MD, FRS
Read Nov. 23 1758

It is in this rock, that the Ammonitae, or Snake-stones, as they are commonly called, are found, which have undoubtedly been formed in the *exuviae* [cast-off shells] of fishes of that shape; and though none of that species are now to be met with in the seas thereabouts, yet they in many particulars resemble the Nautilus, which is well known. The internal substance of those stones, upon a section thereof, appears to be a stony concretion, or muddy sparr. Stones of the same matter or substance, in the shape of muscles, cockles &c, of various sizes, are also found therein, and now and then pieces of wood hardened and crusted over with a stony substance are likewise found in it.

Many naturalists have already observed, that among the vast variety of extraneous substances found at several depths in the earth, where it is impossible they should have been bred, there are not so many productions of the earth as of the sea; and it appears by the accounts of authors both ancient and modern, that bones, teeth, and sometimes entire skeletons of men and animals, have been dug up or discovered in all ages, and the most remarkable for size commonly the most taken notice of. In the first particular this skeleton will most probably appear to have belonged to an animal of the lizard kind, quadruped and amphibious; and as to its size, much larger than any thing of that kind ever met with or found in this part of the world; though, from the accounts of travellers, something similar is still to be met with in many of the rivers, lakes, &c. of the other three . . .

These are the principal facts and circumstances attending the situation and discovery of this skeleton; which from the condition it is in, and from the particular disposition of the *strata* above the place where it is found, seem clearly to establish the opinion, and almost to a demonstration, that the animal itself must have been antediluvian, and that it could not have been buried or brought there any otherwise than by the force of waters of the universal deluge. The different *strata* above this skeleton never could have been broken through at any time, in order to bury it to so great a depth as 180 feet; and consequently it must have been lodged there, if not before, at least at the time when those *strata* were formed, which will not admit of a later date than the above-mentioned.

P.S. In the xlixth vol. page 639, of the *Philosophical Transactions*, an animal is described by Mr. Edwards, which was brought from the Ganges, and resembles this in every respect. He calls it *Lacerta (crocodilus) ventre marsupio donato, faucibus Merganseris rostrum aemulantibus*."

A postscript to Chapman's letter mentions that 'The bones were sent up, and are herewith presented to the Royal Society by J. Fothergill.'

The two Whitby men were, by some luck and some judgement, quite right about the skeleton's taxonomy. It is a marine crocodile, of a species which inhabited the continental shelf waters of north-west

Europe around 185 million years ago. The skeleton was originally classified as a specimen of *Teleosaurus chapmani,* but was subsequently allocated to the species *Mystriosaurus bollensis.* In 1781 the Royal Society donated the skeleton to the British Museum. It was transferred to the Natural History Museum when it opened in South Kensington. A museum guide for 1910 shows that the fossil was on display at that time:

> "The incomplete skeleton of *Mystriosaurus* from Whitby, in Wallcase 3, is interesting from the fact that it is the actual specimen described as an 'Alligator' by Chapman and Wooler in the Royal Society's Philosophical Transactions for 1758."

At present it is in storage in the museum (specimen number BMNH R1088).

chapter four

# A Journey to a Birth

## *William Smith and the invention of stratigraphy*

A hard coming we had of it, to the white-walled city of York that day. It was an ill-tempered start; the wind, mocking our plans for a 'summer' journey, got up out of the north-east and threw rain at us in squalls. The coachman wore out his whip beating the horses on; the animals, not seeing the sense of running into a head wind, sought every chance to duck into the shelter of a copse or a promising looking courtyard. I would have felt sympathy for the beasts, were not our plight so sorry itself. Me and Dr Perkins alternately clinging to each other, and being thrown apart and together by the furious gusts, barely able to speak for the nausea, and unable to hear for the sound of the wind. Deaf-mutes in a padded cell, being shaken as if by the hand of God. One minute on the floor of the carriage, the next apparently stuck to the ceiling. It was a wretched journey, with barely ten miles covered in four hours of travel. And all the time the voices singing in our ears, saying it was folly not to rest, to let the gale blow through, to watch only its fitful disturbance of some inn fire. But the choice was never ours, though it was in truth supposed to be. I know now, as I felt then, that this was not our journey, but that of another, of our 'companion'.

The year was 1794. After much representation, the Act of Parliament licensing the building of the Somerset Coal Canal had been passed that summer. Dr Perkins and myself, as members of the Board, were immediately requested to make a tour throughout England and Wales, to procure the best information we could on canals and on coal mines; and report the same to the company of proprietors. Our journey was to take us through the coalfields of the Midlands, Yorkshire,

Durham and on to Newcastle. We might see how coal is extracted from the earth and how it is transported, in these other parts of the kingdom. We were to be accompanied by Mr William Smith, a young man engaged on surveying work for our company. There was no doubt of the value of his work. Unlike others he had the benefit, or perhaps had taken the trouble, of seeing the different ways in which canals and mines had been laid and excavated. From this, he often said (so often, I confess we tired of hearing it), we would avoid the errors of those who had gone before. Were this all he said and all he thought, we might have been on some 'ordinary' journey – three men going about the normal and eternal business of learning from the errors and achievements of others. But Mr William Smith was not there for this. Or at least not for this alone.

Let me say that Dr Perkins and I were not untutored men. We were educated by our betters in the ways of men, and by our own experience in the ways of nature. The responsibility for this journey was not laid on us lightly. We were, in ways that may seem foolish now, 'expert'. We were well equipped to examine the mine workings, the canal laying, the arrangements of the great engines that were to greet us on our travels. We were men of our time. Lest we suffer in history in comparison with our fellow traveller, I shall say, reader, that we were just like you. Our world was changing before us, and we half understood it. Partly scrabbling to keep up, partly feeling ourselves to be just ahead of a monstrous game in which those who pause are swept away. But often catching ourselves wondering at the speed of it all, and just as often trying not to.

I can only explain it now by saying we were trying to tell ourselves we were part of it all. But a man's lifetime is longer, in many ways, than we imagine. And yours too will be long enough to feel the disappointment that change has brought. They say that this is an illusion that softens the prick of death – that we must feel we no longer quite belong to the world before we quit it. That it has altered away from us. But while there is no real 'golden age' to be rediscovered, so there is no 'dark age' through which humanity has passed, that was not mourned by some who lived through it. This perhaps was our difference. We sensed these things to come in ourselves, but Smith did not. He saw not just that things would be better, but that they would be transformed for ever, and in ways which he knew himself he would not understand.

We set out in high spirits from Bath, each with a memorandum book to record our experiences. Mr Smith insisted on riding at the front of the chaise, where he might see 'the lie of the country'. This suited us well enough, though but for the times he called a halt in order to leap upon some poor unsuspecting rock, we would have travelled at double the speed. We climbed into the Cotteswold Hills in summer heat, stopping at Tetbury and Thames head to inspect the canal works. The great tunnel at Salperton Hill was then complete, running four and a quarter miles at a depth of 90 yards below the surface of the earth. It remains a wonder that men can first dare to conceive and then construct such works. At Kingsnorth on the Worcester and Birmingham Canal, another such tunnel was in progress. While Dr Perkins and myself took note of the engineers' arrangements, Mr Smith spent much time looking at the rocks through which these man-made caverns ran. I confess this was of little interest to me, as they would surely be different from our own rocks in Somerset – and in any case a rock must be dealt with when you come to it. On this the ingenuity and resourcefulness of our engineers had not yet failed us. Mr Smith was, I felt, veering from the practical nature of our task; though when Dr Perkins, in a jocular manner, suggested as much, the young man became very agitated. From then on, we felt inclined to let him at his 'studies'. He was, let it be said, as conscientious as we in the technical matters. These were not secondary to him, but an integral part of whatever task he had set himself.

On through Birmingham and then to see the silk mills at Derby. On the north side of Birmingham we stopped to inspect the tiers of locks passing through the red rock of that region. At Derby and Ripley we saw coal that was unlike that from Somerset, and then went on to Chatsworth and to rest at a comfortable inn at Matlock. The landscape and rocks that make these hills seemed different from our own, though Mr Smith would tell us that the similarities were as important. We saw the tunnel of the Cromford Canal at Butterley Park, where there was later a great ironworks, and then went to the colliery at Hisley Wood, next to the town of Chesterfield (a poor place, though it stands in the midst of rich coal workings). At Hisley Wood, on the estate of Earl Fitzwilliam, Mr Smith and Dr Perkins were lowered into the coal pit on a crossbar attached to a steam-engine, generally used for lowering and raising corves.* As they descended on one arm of the engine, so a full load of corves ascended on the other, and passed them at an enlargement of the shaft, mid-way. Dr Perkins was, I am afraid, unable to understand the workings of this ingenious machine, though Mr Smith, for whom no mechanical device held mystery for long, took the greater part of our journey from Chesterfield to Barnsley in attempting to explain it.

At Leeds we reached the furthest extent of the Yorkshire coalfield. Further north there were no canals. So it was that we agreed, on Mr Smith's suggestion, to strike east to see the great Minster at York, before going north to Durham and Newcastle. The inn near Leeds was poor enough, you may say, with men gambling and calling through the yard half the night. The London coach came in at a God-forsaken hour, held up on the road by weather we had yet to face. And as people will at the end of a long journey, the passengers called loudly for their supper, and made much of their adventure in the rooms below.

* Miners' baskets or tubs for carrying coal.

The innkeeper's late night did not improve his disposition or his breakfast, and we were glad enough to be out of there in the morning. The wind was in the horses' faces, slowing us a little at first. Our driver grumbled as was his habit, but we crossed the flat land of this part of the Vale of York with little trouble. We stopped for refreshment at the small town of Tadcaster, and it was when pulling out from the yard there, that the storms got up. It was then that we might have pulled back, but what could be done? The storm might slip by, we might be better going east than standing still. We, after all, had business to attend to. I did wonder that this, the only day we had deviated from our commercial interests, should pay us so ill.

The road lifted us out on to an elevated area known to the people there as Tadcaster Moor. A more unpleasant place to be in such weather I have never seen. Flat open heathland with not a scrap of shelter. The odd dismal clump of trees, and one or two hovels – though God knows how the residents lived off such land. Mr Smith, as usual, was riding up at the front. I could hear his shouted instructions to the driver being thrown away by the wind. At the point where my stomach was about to spill its contents, the coach suddenly lurched to halt beneath a plump of beech trees. At last, I thought, Smith has made the man understand, we can go no further.

I leaned out of the window to address the others, though the wind in the trees defeated all sound. But then in any case I saw Smith pulling a piece of rock from the ground with the small pickaxe which he habitually carried. He saw me looking and hurried over.

'Mr Palmer!' he had to shout into my ear to make me hear. 'See!' He thrust the piece of rock towards my face.

'Limestone, Mr Smith!'

'Yes! But not like ours. This is magnesian, not calcium limestone. I have heard of it, but never seen it before!'

'Excellent!' It hardly seemed the time or place for a discourse on minerals. The trees might not crash down on us, but their noise altered all perception, so that it would not have seemed remarkable for us all to be swept into the air. 'Do you not think, Mr Smith, that we should turn back?'

I have rarely seen a man look so utterly bewildered. It was if I had said, as I believe I had, Let us put aside our wish to change history, as the weather is inclement.

'But come and see, Mr Palmer, sir, come and see!' He was pulling at the carriage door as he spoke, and then at my elbow, so that I was obliged to follow. I staggered as my legs bowed under my own weight, and would have collapsed to the ground, had not Mr Smith held me up. I felt weak and giddy, and was almost literally carried around the carriage in the man's arms. He propped me against a tree and bade me look up. As I did, with the roaring wind full in my face, I saw a very strange and appealing sight. There, spread out in the middle distance, lying on a flat plain below the moor on which we stood, was a splendid city. The state of my mind, and the curious light that told of the end of the storm, made the city altogether dreamlike. Its pale walls stood out against the dark country. And there, rising above it all were the white towers of the Minster, catching the first of the sun creeping under the clouds from the south, luminescent under the black sky.

'There!' said Mr Smith, the smile of the devil on his face, and lifting his piece of limestone before us, 'I believe we shall see a great deal more magnesium before the day's done!'

An hour's rest at the Black Swan in Coney Street repaired the damage of our journey. We met in the parlour at six and persuaded the landlord to serve us some refreshments. While waiting for Dr Perkins, Mr Smith and I conversed with two gentlemen who were resident in the city. Though there are no mines or canals in the vicinity of York, we were eager to know something of its recent history. Mr Smith, as ever, had little time for social discourse, pressing rapidly on to technical matters. York, it seemed, had been much improved in recent times by two measures which were of great benefit to the population. Firstly, those houses which were built in such a manner as to almost meet in their upper storeys, had been demolished. Sun and air could now reach down into the streets and the lower apartments. Second, the city stands at the confluence of the River Ouse and the River Foss. In previous times these rivers often ran low, leaving great banks of mud and dirt in the heart of the city; also the filth of the common sewers, which they were unable to wash away. The building of locks four miles below the city has been a great advantage. The rivers are kept high, broad and spacious, contributing to the salubrity as well as the beauty of York. The Foss, which was previously a nuisance, has also been made navigable. Mr Smith and myself were much pleased to hear how the application of engineering skills (our own business) had brought such beneficial effects. With that, Dr Perkins arrived, we took some coffee, and wandered out into the ancient streets.

The city, being so old, is made for walking, so that we were able to reach the Minster in a few minutes. The wind had dropped to a breeze, and the sky a pleasant deep blue. The great west front appeared to us as we came on to Duncombe Place. One of the great sights of any man's life, it seems more of its city than the other great cathedrals of the kingdom. We paused to catch the sense of soaring stone, shooting upwards as if from the earth on which we stood, before we ducked out of the noisy street into the dark interior. Dr Perkins and I looked around at the monuments, the tracery and the stained glass, while Mr Smith inspected the building stone with a zealousness bordering on the sacrilegious. We completed our circuit of the central chancel, walking back down the south side. Mr Smith hailed us at the door of the south transept.

'Well, gentlemen, would you fancy a climb before dinner?'

'Where to, Mr Smith?'

'To the top of the tower! This gentleman' – he gestured to one of those somnolent characters who seem to inhabit the dark corners of every cathedral – 'has assured me that the view from the tower has no equal in England. We might see 30 miles in any direction on such a day as this. What do you say? Will you join me?'

'That's right,' the dark figure said. 'It is a hard climb, 290 steps all told. But well worth it.'

There was no mistaking the Yorkshireman's delight in placing this dilemma before us. We both paused, but Smith was not done with his persuasion.

'You might never visit the city again after all.'

'I'll do it!' I said.

'And I,' said Perkins.

'Lead on then, Mr Smith,' I said 'and we shall endeavour to follow.'

The steps began shallow and the going was comfortable. But soon we were into a tight spiral with each succeeding step near as high as my knee joint. Mr Smith's broad frame disappeared into the turns in front of me, and it gave me no satisfaction that Perkins soon did the same behind. The iron hand-rail was replaced by a rope which offered little support. I felt the blood begin to bang in my ears and paused in a doorway. We were less than a third of the way to the top (I was counting the steps) and my legs were advising me against all further upward movement. I looked up at the disappearing curve of the spiral stones, which seemed to press inwards as they circled on. When I peered down I felt a little giddy again, and resolved not to do so. I called down the shaft,

'Dr Perkins. Are you making progress?'

I was relieved to hear the man's footsteps, and then to see his bony hand on the rope. His face was pale and damp.

'Perhaps you had better precede me until we reach the transept roof.' I said. 'We can rest there.'

Frankly, it made me a little stronger that someone else looked so much worse than I felt.

The transept roof was reached and there stood Mr Smith in the open air, annoyingly fresh and willing. He produced his piece of Tadcaster limestone from his pocket, and held it against the wall of the Minster.

'You will have noticed the similarity – '

'To tell you the truth, Mr Smith – ' I gasped for breath as I leaned against the parapet.

*The Central Tower of York Minster*

'It's a steep climb all right,' he conceded, 'but a good opportunity to inspect the fabric of the stone.'

I looked over at Perkins. His head was bowed over the low wall.

'Are you unwell, Doctor?' I asked him.

He took a handkerchief from his pocket and dabbed his face. He turned and started to speak. 'No, no. Just a little exhausted – ' he said, when a look of dismay crossed his face. I followed his line of sight: above us stood the magnificent and terrible sight of the great central tower.

'Oh,' he said, 'I thought we were a little further up than that.'

'Just about halfway,' said Mr Smith. 'Given the view from here, think of the glories to come! And from here we can see only to the south and west. From up there," he pointed to the top of the tower, 'the whole world might be seen.'

'But not by me, I'm afraid,' said the doctor. 'You two go on. I will enjoy the view of the city from this position.'

'And you, Mr Palmer?' said Smith.

'I will go on, but slowly. You must allow me ample time, and remain within my sight.'

'Of course.'

Smith was as good as his word. And while the tower staircase was even steeper than the transept's, we made good time, knowing that this was the summit. We emerged together from the gloom of the stairwell blinking in the blue brightness. It was as if we were within the sky itself. A broad promenade ran around the four sides of the square tower, inside a high parapet, and around a flat lead-covered roof. The storm had cleared any summer haze from the air, which now shone crystal clear, drawing my eyes to the horizon. Smith made straightway to the far side, and looked intently into the distance in a direction I judged to be due east. There was no one else on the roof except myself and my supposed companion. I say 'supposed' because it was clear that, while we occupied the same place, we were as far apart as York and Bath. As I made my way round I called over some inane comment regarding the view of the river. He did not hear me. I continued my circumambulation, passing him on the way, and then made my way round again to his shoulder.

'Something holds your attention, Mr Smith?'

He turned and looked at me with an entirely blank expression.

He had heard my words, but taken in nothing of their meaning, such as it was. We would have stood there an age in mutual incomprehension if I had not tried again.

'Has something caught your attention, Mr Smith?'

Now he looked like a man waking, not from a dream, but from a contemplation so vivid it occupies all one's senses. He smiled.

'Why yes, Mr Palmer. Look.'

I followed his gaze. Flat water meadow and marshes occupied the middle distance, before giving way to pasture land, and then the green billows of a range of rounded hills, lying like so many loaves across the horizon.

All this I saw, and knew that I did not see enough. Strain my eyes as I might, I knew I would never see what William Smith could see that day. What he saw was not in the visible world. He saw it with his eyes, but what he saw was in some other world just beyond the reach of ordinary perception. Though I felt the sense of what he was doing, I did not know then just how out of the ordinary it was. For Smith was not just stepping from the world of the ordinary observer into the world of systematic knowledge, he was in the process of inventing that other world – he was, as we stood 200 feet above the earth, grasping something in nature that was always there, but was apparent to no one before. He was constructing a new science.

'You see those hills, Mr Palmer.'

'I do.'

'You may not know it, but it is possible from their contours, the lack of trees, and their general appearance to say they are of chalk.'

'That is a remarkable skill. I congratulate you –'

'No, no, that is not the point.' His impatience with me was perhaps necessary, and I tolerated it. 'It is the pattern that is everything.'

'I see,' I said, though I confess I did not.

'You remember the canal at Kingnorth and the locks to the north of Birmingham?' He was animated now, as if excited himself by the words that he was saying.

'I remember them well.'

'The rock they passed through was a kind of red marl and sandstone mixed up.'

'I remember you saying so at the time.'

'In each case it lay as an unconformable cover to the coal measures.'

'What exactly does that mean?'

'It lies over them, but there is an interruption between the beds, that is all. Just as at the colliery at High Littlejohn, and all over Somerset.'

'I see.'

'And then we came on to the limestone of the lias, in Derbyshire and here in Yorkshire, just as in the Cotteswold Hills.'

I must have looked a little confused at this, as Mr Smith saw the need to explain further. He grasped my hands and held them flat and horizontal between his.

'The rock strata are formed like this, the oldest beneath and the youngest on top.'

'Yes, I know that much.' I might have been annoyed at him, but his enthusiasm was a pleasant tonic.

'Now, everywhere we have been, the rocks have tended to dip like this –' he twisted our hands slightly – 'toward the south-east.'

'I understand.'

'So that when the top is levelled off, as at the surface of the ground –' Smith slid our hands in their diagonal aspect so that the edges of them made a flat surface – 'the oldest rocks are to the north-west, and the youngest to the south-east.'

We both looked at our hands for a moment or two; then became a little embarrassed and dropped them. Mr Smith looked out across the country, and seemed in danger of re-entering his private world.

'And the chalk, Mr Smith?'

'Ah yes, the chalk, Mr Palmer. The chalk, you see, is the youngest yet. If this pattern is true – if it is repeated all over the kingdom – then to the south and east of the lias limestone, and the red sandstone and the coal, there must always be chalk. And there,' he pointed to the eastern horizon again, as if to keep the vision of the hills alive, 'as on the downs of Wiltshire and Hampshire, is the chalk, Mr Palmer.'

'Then you are to be congratulated on a notable discovery, Mr Smith.'

This approbation was evidently not what Mr Smith wanted.

'No doubt you think me a little strange for this obsession with strata.'

'Not at all – ' I started, but he had little time for polite denials.

'You are a coal-mining man, Mr Palmer, and for that you have my respect. Now if you were to go to those chalk hills and dig for coal, you would never find it. Never.'

Though I had believed I understood Smith's explanations, this seemed a truly fantastic inference.

'But how can you know that, Mr Smith? Those hills are more than twenty miles away.'

'Think on it, Mr Palmer, think on it. You would need to dig through chalk, which is perhaps 500 feet thick, then through shale, limestone and red sand before you even got to the coal measures. You would be miles under the earth.' He laughed to himself at the thought of it.

'Can this be true?' I asked.

He turned to me, his face and eyes set strangely with a conviction that I had never seen on any man.

'Not "Can it be true?" Mr Palmer. It *must* be true. These are laws of nature. They have no variance. What I seek is a philosophy which will embrace the whole earth. It cannot allow any exceptions.'

After an exhausting day we dined exceedingly well at the Black Swan, and indulged ourselves with a pineapple each in reward for our endeavours. From York we went on through the Durham coalfields to the city of Newcastle, from where we returned to Bath via Lancashire and Shropshire.

After that journey I saw little of Mr William Smith. He remained in the service of our company for some years, before a dispute led him to go his own way. The science of geology only began to prosper some twenty years after our journey, when it was taken up by some gentlemen in London who formed themselves into a learned Society. I hardly thought about Mr Smith in connection with this, until I received a letter only this month, from Dr Perkins. He thought that I would like to know that William Smith, as he put it 'the young man who worked for us, and came with us to the North', was presented with the first Wollaston Medal and named 'the Father of English Geology' by no less a person than Professor Adam Sedgwick, President of the Geological Society of London.

It is this news that has brought back the memory of that day in York, a day that has since been called, 'the birth-day of the science of stratigraphy'. In the thirty-seven years since, I have managed my estates and brought myself and my family a good living. That is as much as a man might be expected to do. But I think then of William Smith and his 'philosophy', of how he brought a new world into being – and had to wait half a lifetime for his reward. And then I feel a shame at my contentedness, and I wonder, what are the limits of what any man might do?

Samborn Palmer esq., *October 1831*

chapter five

# The Second Reptile

## *The vicar, the fish-lizard and the question of extinction*

George Young was a great figure in the beginnings of geology in Yorkshire. He wrote the first ever book on the geology of the coast, published as *A Geological Survey of the Yorkshire Coast* in 1822. He was the moving force behind the founding of the Whitby Literary and Philosophical Society and the opening of the museum which the Society owned. He fought hard and passionately for the promotion of the science of geology and for the acquisition of important fossils by the Whitby Museum. In 1819 he was called to inspect a newly discovered fossil skeleton. It was unlike anything he had previously seen. His description was reported to the Wernerian Society in Edinburgh in 1820.

"*Account of a Singular Fossil Skeleton, discovered at Whitby, in February 1819* by the Rev George Young, AM, Whitby, Yorkshire

> The skeleton was imbedded in the alum-rock, where it is washed by the tide, and covered at high water, about half a mile east from the entrance of Whitby Harbour, and ten yards from the face of the steep cliff, which there fronts the German ocean. The cliff at that place, is about sixty yards in height; which, of course, was the depth of this skeleton from the surface, before that part of the cliff, which formerly covered it, was washed away. The skeleton lay in the upper part of the great aluminous bed, which here descends below high-water

mark. Its position was nearly horizontal; the top of the cranium having first made its appearance. This must have been exposed for some years, as it is considerably water-worn; an accumulation of loose stones, over that part of the rock, having both concealed the skeleton and contributed by their rolling to wear its surface.

The skeleton as may be seen in the drawing, is not only in a mutilated state, but dislocated and rent into pieces, indicating some terrible convulsion at the period when it was imbedded."

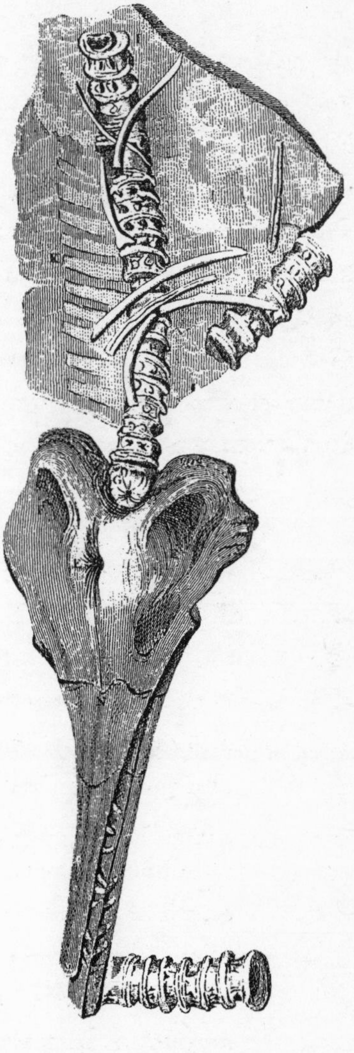

*The original drawing of the fossil described by George Young in 1819. It has since been identified from this drawing as an ichthyosaur of the species* Leptopterygius acutirostris.

A detailed description of the skeleton follows, before Dr Young turns to the problem of identification.

> "To what class of animals this skeleton, and others found at Whitby, should be assigned, it is difficult to determine. They appear to have little or no alliance with the crocodile family. No portion of any crustaceous covering has been found with them, nor any part of bones of the feet. The teeth, indeed, resemble those of the crocodile, but differ from them materially in the regularity of their size and their arrangement. The cranium is totally unlike that of the crocodile, as the writer found by comparing the drawing with a specimen of the latter in Dr Barclay's Museum. Besides, the vertebrae are evidently those of a fish, each being in the form of a concave lens, hollow on both sides. It further appears that the animal has had a pectoral fin, similar to a fossil fin in the possession of Mr Bird, found about two years ago, not far from the spot where the large skeleton has since been discovered . . .
>
> No opportunity has occurred of comparing the fossil fin with that of the dolphin; but the descriptions of the latter, by Cuvier and La Cepede, furnish sufficient ground for regarding them as analogous."

At the start of his career in geology, around 1815, George Young was a pioneer in the subject. But his strict views on the authority of the Bible became increasingly hard to reconcile with the changing ideas of leading geologists, and with the developing view that the Bible was a work of moral and spiritual guidance rather than a documentary history of the

Earth. Quite apart from that, there was one issue in geology that Young and others like him were unable to accept – extinction. The last paragraph in his paper on the 1819 fossil may seem a merely optimistic footnote – it was, though, a deeply held article of faith. Any creature found as a fossil must exist as a living being somewhere on Earth.

> "After all, it must be acknowledged that, in the conformation of the pectoral fin, and other particulars, this fossil animal differs from all living creatures hitherto described. It is not unlikely, however, that as the science of Natural History enlarges its bounds, some animal of the same genus may be discovered in some parts of the world. Brown, in his late Travels in Egypt, could not, among all the fishes of the Nile, identify more than one or two, as corresponding exactly with any of the European fishes; and when the seas and large rivers of our globe shall have been more fully explored, many animals may be brought to the knowledge of the naturalist, which at present are known only in the state of fossils."

The Bible does not say that animals cannot become extinct. Nevertheless the idea of extinction was anathema to many others besides scripturalists. The notion that God and Man were locked in a mutually beneficial relationship was commonplace. The unwinding of this notion was a further step in the demotion of humanity from its privileged place in the natural world. George Young put his arguments against extinction in a book called *Scriptural Geology* published in 1840. His major objection was the absurdity of animals having existed and then died out, before Man had been there to appreciate them.

"Besides, it seems scarcely consistent with the wisdom of the Divine Being, any more than with the declarations of his word, that a succession of creatures, all beautiful and interesting, should occupy our globe throughout long ages, without any intelligent creatures to enjoy the scene, and praise the Creator.

All his works, indeed, praise him; but there is a rational praise, which man alone can render on earth; and it is unreasonable to suppose that, during so long a period, no provision should be made for an object so important."

It is easy to ridicule the views of the past, and the apparent arrogance of the scripturalists. But their belief that human rationality was the pinnacle of God's creation was widely held, and was shared by many natural scientists, who used this idea to promote the application of rationality to all things. George Young should be praised for his enormous energy and enthusiasm for his subject. One observation in *Scriptural Geology* was certainly ahead of its time. The notion that life-forms have been somehow 'improving' with the passing of time has remained powerful until the present day, and has been extremely difficult to dislodge from people's understanding of evolution. Young would have no truck with evolution, but eighteen years before Darwin published *On the Origin of Species* Young saw that all creatures are perfectly organised for their particular functions.

"Some have alleged, in proof of the pre-Adamite theory, that in tracing the beds upwards, we discern among the inclosed bodies a gradual progress from the more rude and simple creatures, to the more perfect and completely organized; as if the Creator's skill had improved by practice. But for this strange idea there is

> no foundation: creatures of the most perfect organization occur in the lower beds as well as the higher."

The 1819 fossil, which George Young accurately described as having some features of a crocodile and some of a dolphin, turned out to be an ichthyosaur, which literally means 'fish-lizard'. The ichthyosaurs became extinct at the same time as their cousins the dinosaurs, 65 million years ago. The 1819 specimen, identified from the drawing as probably *Leptopterygius acutirostris,* cannot now be traced.

chapter six

# Stones from the Sky

"It is easier to believe that Yankee professors would lie, than that stones would fall from Heaven."

Thomas Jefferson, 1807

The gods in their heavens become curious, from time to time, about what is happening on Earth. When their curiosity overwhelms them, they pull apart the dome of the sky so that they can look down, and when they do, a star is released and falls across the sky. So when you see a shooting star you should make a wish, because, at that moment, you can be sure that the gods are watching and listening.

But Fate is a spinner of threads, and the thread of each of us is attached to a star. When any one of us dies, a star falls from the sky. It may be a sign of misfortune for the one who sees it, but not if he says 'May God guide you to a good path', and thereby helps the wandering soul to find rest.

## THE MISUNDERSTANDING OF EUDOXUS

Aristotle resists abstraction. He distrusts mathematics as a way of understanding or describing the world. When Eudoxus draws circles, and circles on circles, to represent the orbits of the planets, Aristotle is interested, but wary.

The apparent complexity of the system is not the problem. Eudoxus has worked from the two central axioms of the cosmos – that the Earth is at its centre, and that all celestial motion is circular – and has developed a brilliant mathematical model. Each planet moves across the surface of a series of three or sometimes four interlocking spheres. The integrity of circular motion is upheld, and the orbits of the planets around the Earth are accurately, if arcanely, plotted. Simplicity of principle gives complexity of outcome.

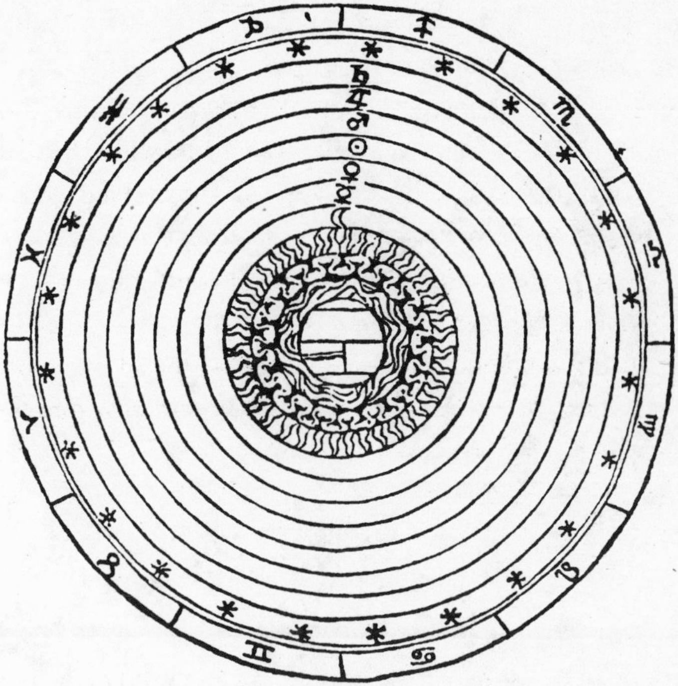

*The universe as described by Aristotle and codified by Ptolemy, from the* Sphaero Mundi *published in Venice in 1490. The planets, the Moon, the Sun and the Earth each occupy their own sphere. Objects cannot move between the spheres, so stones cannot fall to Earth from outside the atmosphere.*

Eudoxus, in his work, makes no physical assumptions. He has the data of planetary observations. He constructs an abstract model and, using his mathematical and geometric skills, he sees how far it will take him.

Eudoxus has unwittingly given Aristotle a perfect model. Aristotle sidesteps the mathematical element of the cosmos of Eudoxus, by simply converting his abstract mathematical spheres into real physical entities – giant transparent crystal orbs carrying the Sun, the Moon and the planets around the Earth. The spheres are like a modern novelty watch, where a ball marking the seconds moves round the watch face without seeming to be connected to anything. In this neat illusion the ball is painted on to a transparent disc the size of the face, and it is the disc that moves. Aristotle needs a Final Cause, an ultimate force; and a mechanism for transmitting that force to all parts of the cosmic system. The force that rotates the cosmos is located at the outermost sphere, the *primum mobile*, or First Motion. The spheres transfer the motion of the *primum mobile* down through the layers of the cosmos. One side-effect is that no physical object can travel through the solid crystal spheres. Stones that fall on earth cannot have come from further than the outer edge of the atmosphere.

The whole thing is absurd of course, but it is necessary. The great historian of science, Giorgio de Santillana, says that Aristotle misunderstood Eudoxus. He does not say whether this misunderstanding was deliberate.

# STORIES FROM THE FIELD OF HEAVEN

## *The discovery of the Campo del Cielo*

In 1553 a group of Spanish soldiers and settlers made the perilous journey across the Andes from Peru, and then turned south. They stopped at a place where the foothills flatten out into plains, on the bank of the Rio Dulce ('the sweet river'). Here they founded the city of Santiago del Estero, the oldest European city in present-day Argentina.

To the north, west and south of their new city, stretching for as far as they then knew, was the chaco. This seemingly endless plain of dry scrub, with no watercourses and no hills to break its numbing monotony, was inhabited by a few indigenous tribes who managed to scratch a living from the unpromising land by gathering honey.

The Spanish had no cause to venture into the chaco, in fact it was considered a dangerous wilderness. But soon after arriving, the soldiers of the garrison began to hear rumours and tales from some of the native people that came to Santiago to trade. Two groups in particular, the Tobas and the Mocovis, told the same story over and over. Deep in the chaco there was a huge piece of metal, the size of a flat wagon, lying half in and half out of the earth. The native peoples said that the metal had fallen from the sky many years previously. For this reason the place where it lay was called Piguem Nonralta, which the Spanish translated as Campo del Cielo, which in English is the 'Field of the Sky' or the 'Field of Heaven'.

Such fables were taken seriously by the Spanish. It was only forty years since Francisco Pizarro had discovered and plundered the unimaginable hoards of gold and silver of the Peruvian Incas. For Europeans, the interior of this strange continent held real possibilities of enormous riches. Explorers dreamed of El Dorado, but men knew that gold and silver did not arise from nowhere – it had to be dug from the earth. The descriptions of the metal at Piguem Nonralta told of a silvery appearance. It was enough for the Spanish Governor of Tucaman to send an expedition.

One morning in July 1576 Capitan Hernan Mexia de Miraval of the Royal Spanish Army led a contingent of eight soldiers out of the garrison fort at Santiago del Estero, and into the uncharted wilderness of Gran Chaco Gualumba. The party faced difficulties on their journey – the barrenness and lack of water in the country and continuous attacks by Chiringuano Indians. But eventually they reached the reported location and found the large mass of metal, which was just as the native people had described it.

Mexia de Miraval decided that the metal, which was deeply embedded in the earth, was the surface outcrop of a seam of iron ('minero de hierro') which ran into the earth. He managed to hack some pieces off the metal boulder and carried them safely back to Santiago del Estero. The samples were shown to a blacksmith, who found them to be of an unusually pure iron.

It seems strange that no further action was taken to exploit this source of pure metal. For whatever reason, and there are many possible explanations, all memory of the expedition disappeared within a few years. A certain don Martin Ledesma Balderrama served as Lieutenant Governor of Santiago del Estero soon afterwards. He took an intense interest in the natural history of the chaco, writing a detailed description of flora, fauna, geography and inhabitants. He even recorded the legends of the peoples of the chacos, including the legend of the iron mass in the Field of Heaven. But he said nothing of Mexia de Miraval, who, it seems, had already been lost to history.

It was another 200 years after Mexia de Miraval before legend and rumour once again reached a level that could not be ignored in the tabernas and plazas of Santiago del Estero. This time the Indian legends were leavened with talk of silver, a rumour that, once whispered, could not be suppressed. A certain don Bartolome Francisco de Maguna set out from the city in 1774. He travelled for 90 leagues (a Spanish league is 4.2 miles) due east, and discovered a great bar ('gran barra o planchon') of metal. This was undoubtedly the same mass that de Miraval had found. De Maguna guessed the weight at 23 tons. He also returned to Santiago del Estero with samples, and this time some were sent to Madrid for analysis. The news that came back from Spain could hardly have been more sensational. The metal from the mass was 80 per cent pure iron, and, astoundingly, 20 per cent silver. Had the fabulous become real?

It is 1776. Settlers in the north of the Americas have rebelled against their European masters, and declared independence from Britain. The Spanish Crown reinforces its hold over its colonies in the South, by establishing a new administrative centre at Buenos Aires. The Viceroyalty of Rio del Plata ('River of Silver') is to have jurisdiction over the southern part of the continent of South America east of the Andes watershed. It will last only fifty years before being overthrown by the founders of the new country of Argentina, 'the Land of Silver'.

In the same year the news from Madrid has reached the Viceroy of Peru and Chile – the most powerful man in South America. He immediately orders de Maguna back to the chaco. Maps are drawn of the location; more samples are taken, and this time sent to the Viceroy in Lima. In Lima, the centre of the continent's silver trade, an assayer disappointingly will only concede that there is some silver in the iron, but does not or will not say in what proportion. The Viceroy of Peru decides to hand over the case of the iron mass in the chaco to his new counterpart in Buenos Aires. He sends all the documents, including the metallurgic analyses of the samples.

The first Viceroy of the Rio del Plata region, don Pedro de Cavallos, is, it seems, a more impetuous man. The information from Lima has sent the city of Buenos Aires into a spin. The place seethes with rumours that the chaco, the mysterious unknown uncharted fearful chaco, contains silver in quantities beyond measure. The chaco is richer than all the silver mines of Peru combined. In a tone of high excitement don Pedro writes to the Foreign Minister in Madrid, don Jose de Galvez, on 22 December 1777. He informs his superior that a silver mine has been discovered at Santiago del Estero; it was previously mistakenly thought to be an iron mine, and was even said by 'the common people' to be a mysterious 'meteor'. If the mine contains only one-fifth of the silver that the metallurgists have said, don Pedro argues, then it will still bring huge wealth. He urgently requests a shipment of mercury be sent from Spain, for the processing of the silver that is to come from the new mine.*

In all the excitement Mexia de Maguna is keen not to be forgotten. On 2 May 1778 he writes to don Pedro de Cavallos to remind him that he was the discoverer of the mine, and to ask that his name be mentioned to His Majesty the King of Spain. He sends yet another sample of material to don Pedro, who asks four master blacksmiths in Buenos Aires to examine it. They testify under oath that the iron is superior to anything they have previously seen or worked. Their report does not mention silver.

On 5 June 1778 the Foreign Minister, clearly impressed by the work of his new Viceroy in Buenos Aires, writes to tell him that 2,000 quintales ('dos mil quintales')† of mercury, sewn into pigskins have been dispatched from the Spanish town of Amaden, presumably *en route* to Buenos Aires via Cadiz. We do not know whether the mercury arrives.

* Mercury was employed in the 'Patio' process of silver extraction, first used in Mexico in 1557, and continuing in use until around 1900. The silver ore was made into a paste by grinding it in water. Salt, copper sulphate and mercury were added, and the mixture was stirred by walking mules through it. The mercury was removed from the resulting silver amalgam, by heating in a retort.

† The historian Ursula Marvin, on whose work this piece draws heavily, says that 'An 18th century quintal was equivalent to 100 pounds; hence it appears that 200,000 pounds (nearly 91 metric tonnes) of mercury . . . were sent across the Atlantic.'

The rank of those controlling the expeditions into the chaco continues to rise through the hierarchy of the Spanish Empire. In the autumn of 1778, Foreign Minister don Jose de Galvez writes to Don Pedro again, this time ordering him to send another expedition to the 'Meson del Fierro' (the table of iron) as the mass has now become known. Before this mission can leave, a man named Francisco de Serra y Canals writes to the Intendente General (a senior government official) in Buenos Aires. He has supervised the analysis of another sample at Uspallata, an old established mining town in the Andes. He writes that there is, in his opinion, no silver in the metal – though the iron is of high quality.

Another expedition marches, with flags flying, out of the gates of Santiago del Estero on 20 July 1779. Led by Major don Francisco de Ibarra, the party of soldiers takes just six days to reach the Meson del Fierro. They measure the mass, as being 3.54 x 1.85 metres in area, standing between 1.36 and 0.84 metres above the ground, but sloping sharply to the south, where its edge is below ground level. They take 2.7 kilograms of samples and return to Santiago, from where the samples are sent on to don Pedro de Cavallos in Buenos Aires. Half are sent to Madrid for analysis, and half examined by a chemist, Dr Miguel O'Gorman, in the city. O'Gorman confirms the growing suspicion that there is no silver in the metal.

By the 1780s hopes of finding vast silver reserves in the chaco had evaporated. The dream of El Argenteo had vanished. For official purposes the King of Spain felt moved to decree that the chaco was the private property of the Spanish Crown. In 1801 extraction of, or trade in, silver from the chaco was forbidden. By that time it was an empty gesture; everyone knew there was no silver there. The Viceroy in Buenos Aires, ever optimistic, ever persistent, sent another and final expedition to the Meson del Fierro in 1783. Though coming at the end of a long line of similar missions, this expedition was to bring enduring fame, and a place in scientific history to its leader. This time the Viceroy was more cautious in his aims, but even in these he was to be disappointed. Was it possible that the iron mass was the tip of a great mountain of iron, or at least the outcrop of a vein of iron ore? A lieutenant in the Spanish navy, don Michael Rubin de Celis, was chosen to lead a force of 200 men. He was to investigate the site thoroughly, and, if there was enough iron to sustain a mine, he was to form a settlement. Though his mission was commercial, he went about his work with scientific thoroughness. It should be noted that, though his conclusions were in error, his observations were invaluable to those who followed.

Rubin de Celis became part of scientific history because he grew inquisitive about the origins of the strange mass of metal; but mostly by the apparently simple, yet astonishingly profound act of placing his samples and the account of his expedition within the world of natural philosophy. De Celis found the Meson del Fierro, measured it, dug around it, levered it up, moved it a little and exploded gunpowder under it. He found that it didn't extend down into the earth and was able to estimate its weight at 15 tons. He wore out seventeen chisels hacking 12 kilograms of sample metal from the mass. Having established that it was not connected to the earth, he began to wonder what it was, and how it got there.

> "Either this mass was produced in the spot where it lies, or it was conveyed hither by human art, or cast hither by an operation of nature. And whence, by whom or how, could it be conveyed hither, as there are no iron mines within hundreds of leagues, nor remembrance that any have been worked in the kingdom? It could be of no value since it could not be used; and why bring it into a country the most uninhabitable of all the Chaco, from the want of water? Besides how could so heavy a mass be conveyed, the Indians never having known the

use of wheel carriage. This mass, therefore, must have been the effect of some volcanic explosion."

There were, though, no volcanoes, either active or dormant, in the region. There was a slight rise in the otherwise flat landscape, and a brackish spring 'about two leagues to the east of this mass'. De Celis noted the ashen appearance of the soil, though in truth it is the same throughout the chaco. He proposed that this was the eroded remnant of a volcano, which had previously ejected the iron mass.

With the return of the expedition of Michael Rubin de Celis, the fantasies of fabulous mineral riches ended, and another story began. The mass was no longer the locus of avaricious dreams, but an object of curiosity. And because men were turning curiosity into a systematic rite, it became a scientific specimen. Instead of sending samples to officials and administrators of the Spanish Empire, or to the crowned heads of Europe, de Celis sent them to scientific societies. In 1788 he sent a letter and a sample to the Royal Society in London. The quotations above are from the letter which is included in the *Philosophical Transactions* of the Society. As well as describing the mass, and speculating on its provenance, Rubin de Celis gives a little historical background. Clearly a man of his time, he is aware of enough natural philosophy to know that certain ideas are 'against the laws of nature' and are open to ridicule. Nowhere in his letter does he mention the Indian legend that the mass fell from the sky, and he does not refer to the site as the Campo del Cielo, the Field of Heaven.

As far as we know, Rubin de Celis was the last man ever to see the legendary Meson del Fierro. Although the Campo del Cielo has been located and investigated in recent years, the original 'Meson' has never been found again.

Europeans have always thought of Rubin de Celis as the discoverer of the Campo del Cielo, while Argentinians have more correctly favoured Francisco de Maguna, who was there in 1774, nine years earlier than his rival. But both choices are wrong. By chance, or by the habits of a bureaucratic state, an account of Mexia de Miraval's 1576 expedition to the Field of Heaven had been written up and signed on 28 February 1584. This document, together with the original orders for the expedition from the provincial Governor, found its way to Spain, where it was placed in the Archivos General de Indias in Seville. This first description of a meteorite in the Americas by a European was to lie undiscovered for 340 years. It was not until 1920 that de Miraval's orders and his own account of his expedition were discovered by a diligent researcher in the Seville archives: the historical claims were wrong by more than 200 years.

But Rubin de Celis is the key figure for the history of science. Although he refused to think the unthinkable, his sample and his letter arrived in London at a time when others were beginning to do so. Within seven years of the receipt of his sample and the publication of his description of the Meson del Fierro, two further events provoked the Royal Society into action. A shower of stones fell on Siena in 1794; and, most spectacularly, in December 1795 a rock crashed from an empty sky into a field near Scarborough in Yorkshire, and was subsequently displayed to the world in a coffee house in Piccadilly. Comparison of the stones from Siena, Yorkshire and the Campo del Cielo showed that they had a common origin, and that stones really did fall from the sky. These particular places, on opposite sides of the world, are united by an extraterrestrial coincidence.

The Campo del Cielo has now been thoroughly searched, mapped and dated by teams of meteorite scientists from the United States and Argentina. Sixty tons of fragments of meteorite plus twenty shallow craters have been found in a fifty-mile-long area or 'strewnfield'. The pattern of distribution, and the content of some fragments, suggests that a large meteorite may have exploded at low altitude. Carbon dating of tree stumps burned and buried by the explosion show the impact happened about 4,000 years ago. The event must have been extraordinarily spectacular. It is known that this area of the continent was inhabited at that time. Ancestors of the native peoples may well have witnessed the explosion of the meteorite and passed the story down the generations.

Enter the main lobby of Alfred Waterhouse's cathedral to Natural History in South Kensington; avoid the distractions of dinosaur skeletons and stuffed mammals on the ground floor and make your way to the stairs. Pause to look down from the first-floor gallery at the scene below, before hurrying on to the front of the building; turn left into the mineralogy display and walk straight ahead. Ignore, for the time being, the sulphates and anhydrides and olivines and agates to left and right. Walk down the centre aisle looking forward. Your eyes will come to focus on a strange and miraculous object. Bulbous, misshapen, gleaming. Run your hand over its smooth surface. A section has been cut across one narrow end and the internal plane polished. It shines like pewter. This is a piece of meteorite from the Field of Heaven. You will never touch anything from as far away as this piece of metal. Not just 12,000 miles from Argentina, of course, but from the asteroid belt 250 million miles away. The name Piguem Nonralta or Campo del Cielo, or Field of Heaven, lay for centuries in the uncertain ground between myth and history. Now we can read it as a poetic description of a piece of natural history. Science may not always be concerned with the dissolution of myth; it can live alongside some other kinds of understanding.

# AN OBSESSION WITH TRUTH

## *The Wold Cottage meteorite*

Next thing, you'll be telling me you have never heard of Edward Topham. Now this I cannot believe. Captain Topham, the King's own 'Tip Top Adjutant'? Captain Epilogue himself! Well, I'll say I'm surprised – and so would he be. By Heavens he would! If he has not come to your attention, it is certainly through no fault of his own. Never was there a man so seemingly determined to live in the public eye – even when he professed to escape from it. Just a few morsels will whet your appetite – but be warned, here is a man who lived ten lives in one.

Not waiting until he left school before making his mark, young Topham led the famous 1768 boys' revolt at Eton; afterwards he lost out to that scallywag Laurence Sterne (yes *him*, can you believe?!) in gaining the living of Sutton-in-the-Forest at York, for which Mr Sterne immortalised him in print; he then took to the Grand Tour and to the army, and was the officer who earned the gratitude of the nation (and the King) by clearing Parliament Square during the Gordon Riots; after dalliances with fashionable ladies of the theatre he took up, and excelled in, writing epilogues; he then founded the most scurrilous and successful paper of his day, and single-handedly established the limits of the law of libel; but he threw all that in and went off to be a magistrate in Yorkshire, only to find that fame pursued him there in the form of a celestial Stone, dropped from the Heavens slap bang on to his land (yes! now you remember); and once he had disposed of that matter (and changed the minds of all the philosophers of Europe), he took up the sport of dog-racing, and bred the most famous greyhound in all of history.

'It cannot be true!' I hear you cry. 'No man could have done as much in a hundred years!' – but Edward Topham did it all in one short of three score and ten. Do not, as they say on Cheapside, take my word for it:

*Edward Topham, painted by John Russell in 1788.*

inspect the documentary evidence for yourself. Then you will see how a man with an unhealthy (I jest, of course) obsession for truth, and a flair for publicity, knocked both Newton and Aristotle into a cocked hat. It is a strange tale indeed.

We turn first to that estimable volume, *Public Characters* of 1804, to find a contemporary account of Topham's life. I fear I will nevertheless have to interject on occasion, as even this colourful biography omits certain wonderful and famous events. I will detain you no further.

"Major Edward Topham is the son of Francis Topham, Esq LLD who was master of the faculties and judge of the prerogative court of York, at which place he resided . . . It was on this gentleman that Laurence Sterne, better known under the name of *Tristram Shandy*, made his first essay in a little pamphlet which he called 'The Adventures of a Watch-Coat'. Here Major Topham, who was then a boy at Eton, was first ushered into the world of literary warfare, from having it stated that his father, who was there held forth as a watchman, '*wanted to cut the parish watch-coat into a dress for his wife, and a pair of small-cloaths for his son.*'

The subject of all this originated, as we have heard, in a dispute with Dr Fountain, the late Dean of York,

> who having neglected to fulfill an engagement made with Dr Topham, engaged Tristram Shandy to turn his breach of promise into ridicule. The best result was that it became the means of first bringing forth into public notice, and afterwards into public admiration, Laurence Sterne as an author, who was at that period a curate in the country, and till then totally unknown.
>
> Major T passed eleven years at Eton, where he was fortunate enough to be distinguished by frequently having his verses publicly read by the master in school, or, as it is there termed, by being *sent up for good*. He afterwards formed one of the numerous band of upper boys who were very severely punished for being engaged in the great rebellion that took place under Dr Foster, then master, who was a great Latinist, a great Grecian, a great Hebraist, and everything but – a man of common sense . . . and of course not qualified to govern the greatest public seminary in the Kingdom."

Oh dear, poor Foster. Lest we underestimate the gravity of the Eton revolt, see the following news item from that great national forum, the *Gentleman's Magazine*:

> "A great number of the scholars belonging to Eton school went off in a body, having taken some offence at the conduct of one of their masters. Most of them have since returned, and the school is restored to its proper discipline."

We return to *Public Characters* and the main narrative.

> "After leaving Eton, Major Topham went as a fellow-commoner to Trinity College Cambridge. Major T remained for four years, long enough to put on what is there called 'an Harry Soph's gown' . . . From Cambridge he went abroad for a year and a half, and afterwards travelled through Scotland. This little tour became better known, as he afterwards gave an account of it in 'Letters from Edinburgh' published by Dodsley. As the work of a stripling they were so well received, that the first edition was soon out of print."

This account fails to mention that while in Edinburgh (a serious claimant for the most learned place on earth at the time) Topham attended lectures on chemistry given by Joseph Black, on physic by William Cullen and on anatomy by Alexander Munro – the three giants of their respective subjects. Now, let us see what the young Topham had to say about truth and jurisprudence; it will enlighten us when we come to matters of natural history. He took an ill view of some Scottish points of law. For example, on the verdict of 'Not Proven' and on majority voting in juries:

> "By all laws, and in all cases, whatever, every man should be presumed innocent till he is proved guilty – proof is either established or not established at all."

On perjury:

> " . . . perjury or swearing falsely on an assize is here punished by confiscation of moveables, imprisonment for a year, and infamy. Even all this is in my opinion too

> small. A man who endeavours to take away the life of another by an oath falsely taken, ought, on conviction, to lose his own."

And on the taking of written statements from absent witnesses:

> ". . . properly, every court ought to hear the witnesses themselves. Every witness does not literally speak the truth, but his countenance always does. It often happens, indeed too often, that a man's looks give the lie to everything he says, and that you read in his face the designed purpose of deceiving you."

But what got Topham's goat most of all was a splendid juxtaposition of terminology, concerning the swearing of oaths:

> "But what is most extraordinary of all is, that they do not bring out this great Oath in all cases; but have, what they call *a little Oath*. My God, Sir, *a little Oath!* What a prostitution of terms!"

Having established that this high-living young man took a serious, even grave, interest in such matters, we return to *Public Characters.*

> "Then he removed to the seat of all joy, in the eyes of a young man, London, and entered into the first regiment of the life-guards, which in the hey-day of the blood may be thought to make that still greater.
>
> There is a principle about some men that never allows them to be quiet or inactive. Major Topham had this principle in full force. He was soon appointed

> adjutant of that corps, and shortly after established as a character in the windows of all the print-shops under the title of 'The Tip-Top Adjutant.' In truth he was a *martinette* of his day and shortly converted a very heavy ill-disciplined regiment into a very good one; in consequence of this he received several commendatory notices from the King and the old general officers of the time."

While quartered in London with his regiment, Captain Topham played a decisive role in the greatest upheaval the capital had seen in decades – one that nearly swept the Parliament and government of the kingdom away in a mad frenzy. In 1780 Edward Topham was an officer in the Life Guards when an angry crowd, under the leadership of Lord George Gordon, marched on Parliament to present a petition against the Catholic Relief Act. Lest you feel that there must be some honour attached to such a demonstration of the people's will, you should know that the mob was protesting against an Act of tolerance for Catholics. 'No Popery' was their sole rallying cry. On the day when Parliament was almost stormed by a mob, Topham commanded the contingent of troops that cleared Parliament Square. His men arrested rioters and took them to Newgate Prison. But this incited further anger among the malcontents and a week of rioting followed in which Newgate, and much else besides, was burnt down. Four hundred people died in the Gordon Riots. For his part in the defence of Parliament, Edward Topham became a nationally famous figure, and was profusely thanked by King George himself.

Let us make a slight diversion into one of the great issues of the day. The war against the American colonists having ended, and the French wars not yet begun, there were a good many young fellows in uniform idling about in barracks with nothing to occupy them but cards and gin. Disputes arose, as they will, out of idleness, and the young men took to impugning and defending each other's honour, simply for sport. One such 'duel of honour' involved our subject Captain Topham, and was recorded in his memoirs by his friend Mr Frederic Reynolds, the renowned dramatist:

> "The meeting took place on the appointed day; Riddell, attended by Captain Topham; and Cunningham by his cousin, Captain Cunningham. Eight paces were measured by the seconds, and they tossed up for the first to fire; which being won by Riddell, he fired, and shot his antagonist. The moment Captain Cunningham received the wound, he staggered, but did not fall. He opened his waistcoat and appeared to be mortally wounded. All this time, Captain Riddell remained on his ground, when, after a pause of about two minutes, Captain Cunningham declared that he would not be removed until he had fired. Cunningham then presented his pistol, and shot Captain Riddell in the groin. He immediately fell, and was carried to Captain Topham's house in Bryanstone-street; where, on the following day, he died. Captain Cunningham, after a long and dangerous illness, recovering, voluntarily surrendered himself to the judgement of the law: he was tried and acquitted."

The result of the dispute was reported in the *Gentleman's Magazine*, for Friday, 2 May 1783:

> "The corpse of Mr Riddell of the horse grenadier guards, was interred in Westminster Abbey. His grave

> is opposite the monument of the poet Dryden. The military procession intended to follow the corpse was prohibited by special order. The corpse was brought on Thursday night in the most private manner to the chancel; but at the interment on Friday it was attended by Lord Townshend, Marquis of Caermarthen, Lord Amherst, Gen. Bulkeley, and two other General officers as supporters of the pall. Ld Macdonald, Mr Topham, and Mr Andrews, were mourners. About 70 officers attended.
>
> Mr Cunningham, who was wounded in the duel in which the above unfortunate gentleman fell, was said to have died the day before of a mortification in his lungs, occasioned by the wound. But the report has been since contradicted."

A pretty picture, is it not? A young soldier being buried in the greatest church in the land? But notice that the military procession was disallowed, and notice our friend Topham among the mourners. Duelling was displeasing to many, whose sentiments were amply conveyed in a letter to the next edition of the same periodical:

> "Mr Urban
>
> The Gothic practice of duelling seems growing upon us every day. The newspapers overflow with recitals of these savage encounters. But I protest it shocked my sensibility when I read the account of the solemn funeral honours with which a young officer, lately slain in one of these premeditated single combats, was interred at Westminster-Abbey. If a military man of very high rank had fallen in the cause of his country, he

could scarcely have been entombed with more soldier-like pomp . . . Let him have descended to the vault of his ancestors amidst the tears of his mourning relations and his sorrowing friends; but when the state had lost the life of a soldier, the voluntary sacrifice of himself and his antagonist in single combat, I can see no one justifiable reason for his pall being supported by a number of general and field officers, who are paid for fighting the battles of their country, not for *revenging to the death* an affront perhaps received during the moments of convivial merriment . . .

To suppress duelling altogether is perhaps impracticable. But surely to have those borne to their graves, who fall the victims of a cruel necessity, in a parading sort of pomp and pageantry of military woe that could with propriety only be exercised at the funeral of warriors and conquerors, is carrying the point of delicacy in matters of honour to a most extravagant height.

Yours &c. A.O.W.

P.S. It may be worth observing, that several of these military mourners appeared *the same evening*, with their crapes, at Ranelagh:

*Bearing about the mockery of woe,*
*To midnight revels, and the public show.*"

Major Topham was not, it is apparent, fully occupied in his military life. He was indeed bored with it. But there were other things to do in London. We return again to the narrative of his life, and the great triumphs to come:

"The Major, however, was not so absolutely absorbed in military tactics as even then totally to estrange himself from literary pursuits. In the midst of his various avocations he wrote many prologues and epilogues to the dramatic pieces of his friends . . . Amongst those that produced the greatest applause on the stage was a prologue spoken by Mr Lee Lewis, in the character of Molière's *old woman*, which had the effect of bringing for many nights together a full house before the beginning of the play – a circumstance in dramatic history somewhat singular; and an epilogue that was afterwards delivered by Miss Farren, now Countess of Derby.

A circumstance happened about this time to the Major, which, as has been said, gave a sort of distinguishing colour to his future life. Mrs. Wells, of Drury-lane theatre, confessedly one of the most beautiful women of the day in which she lived, through the medium of a friend, sent to request him to write her an epilogue for her benefit. He naturally did not deny her request, and of course the reading and instructing her in the delivery produced interviews which the company of a woman so beautiful must always make dangerous. There are, as Sterne says, "certain chords, and vibrations, and notes that are correspondent in the human feelings, which frequent interviews awake into harmony." What did occur may be easily supposed: a mutual intercourse, in consequence of mutual affection, in progress of time took place betwixt them."

How delicately advised! But it was not enough for a man such as Topham to show his affections in private. He took it upon himself to advance Mrs Wells's public reputation, and this, as the anonymous guide to his biography tells us,

> "gave a new spur to his mind, and a fresh activity to his genius. It was the idea that first inspired the thought of establishing a public print. It has been said more than metaphorically, that 'love first created *The World*.' Here it was realised. Gallantry began what literature supported and politics finished. It was thus, as we understand, from a wish to assist Mrs Wells in her dramatic life, that the paper of *The World* first originated, and which, beginning from a passion for a fine woman, attracted to itself shortly afterwards as much public notice as ever fell to the share of a daily, and consequently a very fugitive publication . . . In one week the demand for the *World* exceeded that which had been made in the same time for any other newspaper."

From the following narrow aside, we can infer the tastes of the reading public:

> "It is a singular fact that the correspondence of two boxers, Humphries and Mendoza, raised the sale of the paper to a higher degree than all the contributions of the most ingenious writers. It was the fashion of that time for the pugilists to send open challenges to each other, and thus publicly announce their days of fighting. This they did through the World, as considering it the most fashionable paper; and their writing beat Sheridan all to pieces."

But more to the point,

> "Major Topham's wishes, therefore were fully gratified. The paper of the World, of which he was editor, had extended itself beyond his utmost expectations. It was looked to as a repository for all the best writers of the day; it gave the *tone* to politics, and, what to him was still dearer, it contributed to the fame of the woman he loved.
>
> But, alas! the dearest and most sanguine of our hopes are but as a breath. Mrs Wells, in her eagerness to appear in a particular part, to oblige the manager of Covent Garden, too soon after a lying-in of her last child, produced a revolution of her milk, which afterwards flew to her head, and occasionally disordered her brain . . . The World, at which Major Topham had laboured for nearly five years lost in his eyes all its charms . . . in a short time he resolved to dispose of it altogether."

But here we must allow another voice to enter. A voice which will, by its different perspective, add a little leavening to this story. Mrs Wells recorded her own view of her relationship with Captain Topham in a book of memoirs. She does not remember him with great affection – unsurprising when we learn that she was incarcerated for madness and indebtedness at the end of their 'affair'.

> "I was captivated with the beauty of his mind; he made me an offer of his hand; but, as we could not legally be united in this kingdom, he proposed going to Italy. I listened to his offer, and as at the time I believed him

> sincere . . . I accepted it. ['At the time' – a telling phrase is it not?] But fortunately for him, the commotions on the Continent prevented our immediate departure, which, from constant procrastination was at length obliterated from the tablet of his memory. We resided in Bryanstone-street for three years, where I became the mother of two lovely children. Thence we moved to Beaufort-buildings, where he established a newspaper called The World. I have, in the course of conversation, often heard the expression, *seen a great deal of the world*; but for my part, I saw *too much of it* – for the principal burden of carrying it fell at last upon my shoulders, as the following letters of Mr Topham to me will prove."

Mrs Wells reproduces a series of letters which purport to show that Mr Topham spent his time touring the country, and issuing instructions to her in London for the running of the *World*. Just one will be sufficient here – Mrs Wells has just given birth to a son . . .

> "My Dear Pud,
>
> I wish you joy on all being over, and on a boy – this is right. Pray take care of yourself, and when you have been quiet some time, *take care of* THE WORLD.
>
> I shall be glad to hear of your being recovered.
>
> Yours ever, (the post going)
>
> E.T."

> "So necessary did he [Topham] conceive my exertions to the existence of *The World*, that he promised me a share in it, which like our *Italianate marriage*, he soon forgot."

As well as all this, Mrs Wells felt she had to endure the transfer of Topham's affections to another woman. Mrs Wells spent some days in Calais avoiding the attentions of her creditors (a sadly common resort in these times), before returning to London.

> "On my arrival in London, I sent to Mr Topham to inform him I was come, when he requested Mr Reynolds to leave me at an hotel for the night. At our interview next morning, the alarm of creditors was again thundered in my ears, and I was persuaded by him to leave London under that impression, when in fact it was for the express purpose of having me out of his way, as he had lately taken into the house a Mrs. Lambert."

Alas, when lovers part, charity quickly flies the scene. Perhaps we should avert our gaze and return to the matter in hand. It must be said that a famous libel case brought against him may have aided Mr Topham's decision to quit the newspaper business. His paper had printed some indelicate sentiments about Earl Cowper, not very long deceased. His relatives sued for libel, but on appeal the courts found in favour of Captain Topham and ruled that a dead person cannot be libelled, and that his relatives must prove the criminal charge of incitement to breach the peace. The case stands as a precedent in English law, thanks to Captain Topham's courage in standing up to his accusers. We might now allow our *Public Characters* biographer to say something apposite about the life of a newspaper editor:

> "They who have known what the daily supply, the daily toil, the daily difficulty, the hourly danger, and the incessant tumult of a morning paper is, can alone know

that chaos of the brain in which a man lives who has all this to undergo. Terror walks before him: fatigue bears him down: libels encompass him, and distraction attacks him on every side. He must be a literary man, and a commercial man: he must be a political man, and a theatrical man; and must run through all the changes from a pantomime to a prime minister. What every man is pursuing, he must be engaged in; and from the very nature and "front of his offence," he must be acquainted with all the wants, the weaknesses, and wickedness, from one end of London to the other.

To view all this might gratify curiosity for the moment; but to live in it is to guide a little boat in a storm under a battery of great guns firing at him at every moment; but even this has advantage; it may endear retirement or make seclusion pleasant. In fact, and without a pun, on quitting the World, Major Topham retired to his native county, has lived two hundred miles from the metropolis, without once visiting it during the space of six whole years.

Who could have done this? Who could have thought that remote hills, solitary plains, and, what is worse, *country conversation*, would have found charms sufficient to detain a *town-made man* from the *streets of London*? The physicians would answer, 'cooling scenes are the lenitives of fever.' After the long labours of a sultry day, where can the weary fly better than to the shade . . ."

Mrs Wells had this to say on the matter:

> "Mr T. sold his paper – as my state of mind did not allow me to conduct it any longer – and was in Yorkshire, revelling in the charms of Miss Walton."

But now, just as the sun appears to be setting on a career more accomplished than any man might rightly expect, comes Captain Edward Topham's finest hour. Everything he has so far achieved appears but a prologue to the main drama. Let us begin this remarkable episode with a letter from Sir Joseph Banks, President of the Royal Society, to Sir William Hamilton, His Majesty's Ambassador to the Court of the King of Naples, and the world's leading authority on the great volcano of Vesuvius. There has recently been a report of a shower of stones falling on the fashionable and historic town of Siena in Tuscany. On the day previous to that remarkable event, Vesuvius had one of its most spectacular eruptions in recent times. Banks wishes to know if the two are connected. We include their small talk, as a token of civilised men.

> "Sir William Hamilton, Naples
>
> Soho Square<br>
> Augt 10 1794
>
> My dear Sir William<br>
> I thank you for your favour of the 5th retd. and the intelligence it contains which is valuable to me because I can vouch for its authenticity. We have had some silly stories told here which you by guess have enabled me to contradict with effect.
>
> Your intended account will be valuable treasure indeed as no man since the Creation has I am convinced been present at such an Eruption endowed with half the

volcanic experience you possess. We shall not meet at the Royal Society till November so I pray you do not hurry yourself in finishing it.

Thank you for your Sonette I am as you a great friend to the natural good sense & unaffected good humour which always abounds in Clever Blackguard I have puzzled at the original till by a comparison with other Italian & with your notes I fancy I understand it & taste at least a part of the original humour You have however left out one line which please to send me by your next It is that in the third Stanza which I deem from your notes ends with Compa' Sebeto

I enclose you a letter from our odd Bishop Lord Bristol It is evidently intended for you as appears from the whole of it, tho it is as certainly directed to me – possibly you have a letter from him intended for me & addressed to you. The story seems too marvelous for belief but as nothing is impossible & you will have an opportunity of enquiring into particulars I will suspend my infidelity till I hear your opinions. If the Stones were realy impelled by a Force which drove them so high as to be 18 hours in their ascent & their descent I conclude they must have passed far above the limits of our atmosphere & if the collected mass of Electricity in the regions above which exploded as soon as they arrived into the dense regions of our Exhalations it will open a new field of discussions which cannot fail to amuse our Philosophers very much if it does not instruct them.

It is my custom always to date my letters from London because I wish all the answers to be addressed

> to that place. I am however realy in the Country & so occupied by the vacancy which has taken place in the County of Lincoln by Mr Pethom [?] being made Peer that I can realy do nothing else but labor to keep the peace of our vast county Accept that as the real excuse for a short & a stupid letter. Give my sincere wishes of all Earthly Prosperity to your Emma & believe me as I am
>
> Faithfully & affectionately yours
> Jos. Banks"

You will remember that Sir William has the good fortune to be wedded to Lady Emma, a woman, it is said, of immense personal charm. You should know too that philosophers at this time held the notion of stones falling from the sky in utter ridicule. It was a superstition which could only be entertained by the most credulous and ignorant. After all, both Aristotle and Newton (a fearsome combination) had spoken against the possibility. Such stones as had been seen falling must come from volcanoes, and those that were found must be formed by Electricity, or Lightning or by Exhalations from the Earth. The letter from the Earl of Bristol, Bishop of Derry, dated 12 July 1794 from Siena, which Banks sent on to Hamilton, contains the following description:

> "In the midst of a most violent thunder-storm, about a dozen stones of various weights and dimensions fell at the feet of different people, men, women, and children; and the stones are of a quality not found in any part of the Siennese territory; they fell about 18 hours after the enormous eruption of Vesuvius, which circumstance leaves a choice of difficulties in the solution of this

> extraordinary phaenomenon: either these stones have been generated in this igneous mass of clouds, which produced such unusual thunder, or, which is equally incredible, they were thrown from Vesuvius at a distance of at least 250 miles; judge then of its parabola. The philosophers here incline to the first solution. I wish much, Sir, to know your sentiments. My first objection was to the fact itself; but of this there are so many eye witnesses, it seems impossible to withstand their evidence, and now I am reduced to a perfect scepticism."

In his reply to Banks, Sir William concedes the involvement of Vesuvius, though in an ingenious way.

> "I mentioned to his lordship another idea that struck me. As we have proofs during the late eruption of a quantity of ashes of Vesuvius having been carried to a greater distance than where the stones fell in the Sanese territory, might not the same ashes have been carried over the Sanese territory, and mixing with a stormy cloud, have been collected together just as hailstones are sometimes into lumps of ice, in which shape they fall; and might not the exterior vitrification of those lumps of accumulated and hardened volcanic matter [have] been occasioned by the action of the volcanic fluid on them?"

Next, our gaze alights on an extraordinary letter published in the *London Chronicle* on 7 January 1796:

"*Extract of a Letter from Sheffield, Jan. 1*

On Sunday the 13th ult. at three in the afternoon, the inhabitants of Woldnewton in the county of York, and the villages adjoining for eight miles round, were very much terrified by a strange phenomenon: A report was first heard, resembling the discharge of two large cannons, one following the other about the space of half a minute, and immediately after a rumbling noise: its direction seemed from East to West: at the same instant a stone fell out of the air, weighing 55 lbs. 200 yards from Wold Cottage near Woldnewton, the residence of Capt. Topham, and not more than 30 yards from three of his servants, who were amusing themselves in the field at the same time; by the velocity of its fall it penetrated the ground 18 inches; it was warm when it fell, the outside very black, and smelled strong of sulphur; immediately followed a very heavy shower of rain. There are many neighbouring gentlemen have been to see it, who probably by this time may have discovered the reason of so unusual a thing; some people supposed it had probably been done by art, but most think it to be a phaenomenon: it is of the nature of the free-stone,* and has shining particles in it when broken, it rings when struck, like lime-stone, and is not any way different in weight or size from our common free-stone."

The location was none other than that piece of land to which Captain Topham had so recently retired. Though he was in London at the time,

* Free-stone generally means the local naturally occurring stone used in building.

he duly sped northwards and took it upon himself to investigate this occurrence for himself, and for the benefit of the world. On 8 February 1796, he wrote from Wold Cottage, Yorkshire (a tranquil spot not a dozen miles south-west of the great resort of Scarborough) to James Boaden, Esq. of the *Oracle* newspaper. The letter was published on 12 February 1796.

> "The very singular phaenomenon which took place near my house in Yorkshire on Sunday, Dec. 20, 1795 [corrected to Dec. 13 in later copies], has excited general curiosity. Being in London at the time, it was impossible for me to know more of it than from some vague accounts in provincial and London papers; and to be certain, from private letters that such an event had happened, on my return here I found that, for a space of nearly three weeks, 30 or 40 persons on each day had come to see the STONE which had fallen; and I found likewise a number of letters from all parts of the kingdom, requesting to me to give an account of the circumstance.
>
> The following detail, which you are welcome to make public, will be, I hope, satisfactory on the subject:–
>
> The exact weight of the Stone which fell, on being dug up, was, by MERLIN'S balance, 3 stone 13 lbs on being measured: it had buried itself in 19 inches of soil, and, after that, in 6 inches of solid chalk rock, whence it was some little time in being extracted. When taken up it was warm, and smoked. At the time it fell there was a labourer within 9 yards, and a carpenter and groom of mine within seventy yards. The labourer saw it coming

down at the distance of about 10 yards from the ground. As it fell, a number of explosions were heard by the three men at short intervals, about as loud as a pistol. The Stone is strongly impregnated with sulphur, and then smelt very strongly. The general texture of the Stone is that of grey granite, of which I know of none that may be called natives of this country.

What renders this the more extraordinary is, that the day was a mild, hazy day; a sort of weather very frequent in the Wold hills, when there are no winds or storms; but there was not any thunder or lightning the whole day.

It fell about three in the afternoon; its course, as far as I can tell from different accounts, was from the south-west.

At Bridlington, and at different villages, sounds were heard in the air, which the inhabitants took to be the noise of guns at sea; but at two adjoining villages, the noise was so distinct of something singular passing through the air towards my habitation, that five or six people came up, to see if anything extraordinary had happened to my house or grounds.

In burying itself in the earth it threw up a greater quantity of soil than a shell would, and to a much greater extent. When the labourer recovered from the extreme alarm into which the descent of such a Stone had thrown him, his first description was, "that the clouds opened as it fell, and he thought HEAVEN and EARTH were coming together!"

From all the various persons who have been to inspect this curiosity, and who are daily coming from different parts, no satisfactory conjecture has yet been hazarded whence it could have come. We have no such Stone in this country: there has not been any where in these parts any eruption from the earth. From its jagged and angular form, it cannot come from any building; and as the day was *not* tempestuous, it does not seem probable that it can have been forced from any rocks, the nearest of which are those of Flamborough-head – a distance of 12 miles.

The particulars of this event are now before the public. I have taken due care to examine the accounts given by different persons, who all agree upon the subject; and from what I have seen, I have no doubt of the veracity of the relations. These count for so extraordinary an appearance I leave to the researches of the PHILOSOPHERS.

I have the honour to be
Your obedient servant
EDWARD TOPHAM"

None, surely, could doubt the word of Edward Topham. But soon there was no need to travel to the outer reaches of Yorkshire to witness the extraordinary phenomenon. Topham arranged with a Mr Bohnen that the very stone should be exhibited in the heart of the capital.

EXTRAORDINARY PHENOMENON.

THE Public are reſpectfully informed, that the STONE of immenſe ſize and weight which fell from the Atmoſphere in December laſt, in the Grounds belonging to Captain Edward Topham, Eſq. in Yorkſhire, which was made known by him through the Public Papers in February laſt, is now, by his permiſſion, exhibited for the inſpection of the Curious, and the Public in general.—Admiſſion 1s. at No. 2, oppoſite the Glouceſter Coffee-houſe, Piccadilly; where will be given, gratis, an exact Engraving of the Stone, with Copies of the Teſtimonials (which will do away any doubt as to the truth of ſuch an Event) with an Account alſo of the PHENOMENON in the ſurrounding Atmoſphere, which attended the fall of this remarkable STONE; which is ſtill more extraordinary, from being impregnated with ſolid particles of various metals, diſcernable to the naked eye. Beneath are the names of thoſe who teſtify to the truth of ſuch a circumſtance having happened; the originals of which will be ſhewn at the Place of Exhibition.

TESTIMONIES ſigned by

| | | | |
|---|---|---|---|
| Capt. E. Topham, | Yorkſhire, | Mr. Js. Watſon, | Yorkſhire, |
| The Rev. Mr Preſton, | Ditto, | Mr. Luke Wilſon, Linen-draper, No. 320, Holborn, London. | |
| Mr. John Shipley, | Ditto, | | |
| Mr. Geo. Sawdon, | Ditto, | | |

N. B. The neighbourhood of that part of Yorkſhire where the Stone fell, all agree as to the Phenomenon in the Atmoſphere which attended this remarkable event.

*Announcement in* The Times, *7 July 1796, of the exhibition of the Wold Cottage stone.*

Recall, if you will, Captain Topham's earlier pronouncements on the truth of witness accounts. You may be sure that the following named were careful to avoid embellishment or exaggeration before this fearsome magistrate. We can confirm the rumours that he had constructed several prison cells adjacent to his house to incarcerate those who fell below the standards of the law. Each visitor to the Exhibition of the Stone was provided with a copy of these Testimonies.

"John Shipley, husbandman, deposes, that he was within eight or nine yards of the stone when it fell, saw it distinctly seven or eight yards from the ground, and then strike into the earth, which flew up all about him, and which alarmed him very much. In falling, sparks of fire seemed to fly from it. On recovering from his confusion, he went up to the place in company with George Sawdon, carpenter, and James Watson, groom to Capt. Topham, and helped to dig the stone from the rock of lime-stone where it was stuck. It was buried about 21 inches deep. It weighed, on being carried into the house, about 56 lb. He had ploughed over that very ground the last year.

(Signed) John Shipley"

"George Sawdon, carpenter, above-mentioned, deposes, he was walking with James Watson at the time, heard noises in the air like the reports of a pistol, and was about 50 yards from the place when the stone fell; was certain that there was no lightning at the time. Went up to the spot, and there saw the stone sticking in the chalk which the above John Shipley saw fall. Helped to dig it out, and assisted in weighing it in Merlin's balance – weight about 56 lb. It smelt very strongly of sulphur on being dug up.

(Signed) George Sawdon"

"James Watson deposes to the above effect – has heard different persons in the neighbouring villages mention the noises they heard in the air at that time like the noise of guns at Sea.

(Signed) James Watson"

"I hereby certify to the public, that, while I was in Yorkshire, near Capt. Topham's grounds, I heard noises in the air, like the report of a cannon at a distance; and, at the same time, I felt two distinct concussions of the earth, which shook the buildings and the church near the spot where I was at the time. I was very much surprised, not knowing from what such circumstances could arise. Within a very short space of time afterwards I was informed that a stone had fallen within 200 yards of me and a servant belonging to my uncle, Mr William Park, who resides near Capt. Topham, was one of the people who saw it fall. The above account I declare to be true, as witness my hand,

(Signed) Luke Wilson
Linen-draper, 320, Holbourn"

"Charles Prestin, son of the Rev. William Prestin, 11 years of age, being in the church-yard at play, on Sunday, Dec. 13, 1795, at half past three in the afternoon, after hearing a noise as of firing of cannon, heard at the above time a hissing in the air, and was sure that something fell near the cottage belonging to Capt. Topham. Given under my hand this 29th of April, 1796.

(Signed) William Prestin
Curate of Wold Newton, Yorkshire"

During the exhibition of Captain Topham's stone, which the philosophers as well as the merely curious were bound to attend, the following advertisement appeared in the *London Chronicle*, issue of 18–20 August 1796:

> *This day was published,* in 4to. Price 2s. 6d.
> REMARKS concerning STONES said to have fallen from the
> CLOUDS, both in these Days and in ancient Times.
> By EDWARD KING, Esq. F.R.S. and F.A.S.

In this remarkable document we find the first expression of a thought which was soon to enter the minds of many others.

> "how much resemblance, on the whole, it [a stone from the Siena fall] really has with the stone said to have fallen in Yorkshire; in its gritty substance, and in its metallic and pyrritical spots and in those parts where the pyrites appear to have been decomposed."

But as the evidence of fallen stones became more irresistible, voices were raised against. The celebrated French scholar Guillaume Deluc believed it was all a peasants' fable, dreamed up or imagined by a credulous labourer. Should the fate of the solar system turn on the word of a common ploughman?

> "The York County stone, weighing 56 pounds, which was the initial source of all the hypotheses and all the discussions that have appeared on this matter, had as its basis the report of a labourer, who being in his field, says that he saw it when it was seven or eight yards from the ground."

Should we implicate a whole nation in the obstinate prejudice of one man? When he is French, the temptation is confessedly high. These contrary views were passed to Captain Topham by Mr Bohnen, the exhibitor. That honest gentleman reacted furiously to the imputation.

> "To Mr Bohnen, Piccadilly, London
> Sir,
>
> You mention to me the doubts that some persons have entertained, of the reality of a Stone falling from the Clouds near my house. Those who have read the treatise of the very learned Mr. King, on the existence of such events from the earliest times, may perhaps find their doubts diminished. For my own part, I neither have the leisure nor the inclination to wish to force an imposture on anyone; nor indeed would I have hazarded my name in the first instance, on such an occasion, without the most minute and most accurate inquiries. The accounts have been taken from those who were on the spot at the moment, and who can have no benefit from deceit. To the end of time, it can be of no use to begin to propagate such a falcity; but as it is your request I have taken the Affidavits of those who are now near me; and I do this, from this motive only, that having given you leave to exhibit the Stone for your own emolument, and to gratify the curiosity of the Public, I should be sorry if you were supposed capable of abusing the public confidence. The subjoined relations therefore have been sworn before me as a Magistrate; tho' without Oaths, I was long ago convinced that as the Witnesses to the circumstance had no motive for impolition, they would never have

attempted it with me. Were anything wanting to confirm their accounts, it might easily be obtained by the concurrent testimonies of those who all around this neighbourhood were hearers of the very singular tumulus that troubled the air, at the time the Stone fell; and who were too distant from each other to have fabricated or agreed in a Story; and who are alive to contradict any imposition that might be attempted.

You are certainly at liberty to publish this if you please, tho' I feel a repugnance in again publickly writing on a subject where I can have no human interest whatever, and where my mind is completely satisfied; you Sir, have to answer for this intrusion. One trouble more only shall I give myself; that of erecting a Pillar on the spot where the Stone fell, to perpetuate to those who may view the place hereafter, where such an event took place. If therefore I could guarantee an imposition, I must be weak indeed; for I should have to encounter the observations of my neighbours, and to perpetuate my credulity to posterity. But I have seen enough of the world, not to be so absurd.

I have the honor to be
Sir
Your obedient humble Servant
Edward Topham

Wold Cottage
Oct.17, 1796

And now, the philosophers are finally awakened from their 2,000-year slumber, to investigate with proper exactitude the resemblance of different stones 'said to have fallen on the Earth'. Sir Joseph Banks, having paid his shilling to see Captain Topham's stone, and having been sent some from Siena, and perhaps prompted by Mr King's remarks and his memory of Rubin de Celis's report from the Campo del Cielo, orders an investigation. The first we know of it is at the presentation of the Copley Medal for 1800, the Royal Society's highest annual honour, to the young chemist Edward Howard. Sir Joseph tells the audience that Mr Howard is 'now employed in the analysis of certain stones, generations in the air by fiery meteors, the component parts of which will probably open a new field of speculation and discussion to mineralogists as well as meteorologists'. The time is right – the science of analytical chemistry has recently advanced apace – and Howard is the man for the task. His report, published in 1802, changes the solar system for ever. He describes Captain Topham's stone at length, and becomes, as philosophers are wont to do, a little sniffy about its exhibition:

> "In 1796, a stone weighing 56lbs. was exhibited in London, with several attestations of persons who, on the 13th of December, 1795, saw it fall, near Wold Cottage, in Yorkshire, at about three o'clock in the afternoon . . .
>
> The exhibition of this stone, as a sort of show, did not tend to accredit the account of its descent, delivered in a hand-bill at the place of exhibition; much less could it contribute to remove the objections made to the fall of the stones presented to the Royal French Academy. But the Right Hon. President of the Royal Society, ever alive to the interest and promotion of science, observing the stone so exhibited to resemble a stone sent to him as one of those fallen at Siena could not be misled by prejudice: he obtained a piece of this

> extraordinary mass, and collected many references to descriptions of similar phenomena . . ."

We are here, you understand, at the dawn of a new age, when new natural wonders are not to be advertised and exhibited as freak shows, but soberly analysed and classified. The world prestige of the Royal Society enables Mr Howard to have samples from Siena, the Campo del Cielo in Argentina, Benares in India, Bohemia, Senegal and Siberia (the famous Pallas Iron) as well as the Wold Cottage. He finds undeniable similarities in their composition:

> "It will appear, from a collected view of the preceding pages and authorities, that a number of stones asserted to have fallen under similar circumstances, have precisely the same characters. The stones from Benares, the stone from Yorkshire, that from Sienna, and a fragment of one from Bohemia, have a relation to each other not to be questioned.
>
> Ist. They have all pyrites of a peculiar character.
>
> 2dly. They have all a coating of black oxide of iron.
>
> 3dly. They all contain an alloy of iron and nickel. And,
>
> 4thly. The earths which serve to them as a sort of connecting medium, correspond in their nature, and nearly in their proportions.
>
> Moreover, in the stones from Benares, pyrites and globular bodies are exceedingly distinct. In the others they are more or less definite; and that from Sienna had one of its globules transparent. Meteors, or lightning, attended the descent of the stones at Benares, and at Sienna. Such coincidence of circumstances, and the

unquestionable authorities I have adduced, must, I imagine, remove all doubt as to the descent of these stony substances; for, to disbelieve on the mere ground of incomprehensibility, would be to dispute most of the works of nature.

Respecting the kinds of iron called native, they all contain nickel. The mass in South America is hollow, has concavities, and appears to have been in a soft or welding state, because it has received various impressions.

The Siberian iron has globular concavities, in part filled with a transparent substance, which, the proportional quantity of iron excepted, has nearly the impression of the globules from Benares.

The iron from Bohemia adheres to earthy matter studded with globular bodies.

The Senegal iron had been completely mutilated before it came under my examination."

Mr Howard has the native caution of the analytic man, but he makes clear his thoughts nonetheless:

"From these facts, I shall draw no conclusion, but submit the following queries.

1st. Have not all the fallen stones, and what are called native irons, the same origin?

2dly. Are all, or any, the produce or the bodies of meteors?

And, lastly, Might not the stone from Yorkshire have formed a meteor in regions too elevated to be discovered?"

As soon as Mr Howard's paper was read to the Society, all doubts evaporated. Stones undoubtedly did fall from the sky, and their origins were undoubtedly similar and, it seemed, from outside the Earth's atmosphere.

Sir Joseph Banks died in 1820, ending a reign of no fewer than forty-two years as President of the Royal Society. His successor Humphry Davy gave an address on taking the chair at his first ordinary meeting as President which put the official seal on the new view of 'stones from the sky':

> "When the boundary of the solar system was enlarged by the discovery of the Georgium Sidus, and the remote parts of space accurately examined by more powerful instruments than had ever before been constructed, there seemed little probability that new bodies should be discovered nearer to our earth than Jupiter; yet this supposition, like most others in which our limited conceptions are applied to nature, has been found erroneous. The discoveries of Piazzi, and those astronomers who have followed him, by proving the existence of Ceres, Pallas, Vesta and Juno – bodies smaller than satellites, but in their motions similar to primary planets – have opened up to us new views of the arrangement of the solar system. Astronomy is the most ancient, and nearest approaching to perfection of the sciences; yet, relating to the immensity of the universe, how unbounded are the objects of enquiry it presents! And, amongst them, how many grand and abstruse subjects of investigation! Such, for instance, as the nature of the systems of the fixed stars and their changes, the

> relations of cometary bodies to the sun, and the motions of those meteors which, in passing through our atmosphere, throw down showers of stones; for it cannot be doubted that they belong to the heavens, and that they are not fortuitous or atmospheric formations; and, in a system which is all harmony, they must be governed by fixed laws and intended for definite purposes."

Edward Topham also died in 1820 after, we understand, an active retirement. He took no further apparent interest in natural philosophy; instead, according to *Public Characters,* turning to legal work and country sports:

> "Major Topham, we understand, has not found, even in retirement, time hanging heavy on his hands. The duties of a county magistrate, in a large county, are very great, and very incessant. He has a considerable farm of some hundred acres under his own management, and his occasional hours he is dedicating to the compilation of a *History of his own Life*. Major T. living in the wilds of Yorkshire, among other country amusements, has been the founder of many *coursing establishments*. His greyhound, the famous Snowball, is well known to the whole kingdom, as his breed has been sought after in every part of it."

This is a gross underestimate, as we can say with confidence that Snowball (a black dog, wouldn't you know) was the most famous greyhound that ever lived. On another matter Topham was, as ever, as good as his word. He promised to erect a monument on the exact spot where the

Wold Cottage meteorite fell. It stands there still, among the low chalk hills, bearing the following inscription:

Here
On this spot, *December 13th, 1795,*
fell from the Atmosphere
AN EXTRAORDINARY STONE
In Breadth 28 Inches
In Length 30 Inches
and
Whose weight was 56 pounds
THIS COLUMN
In Memory of it
was erected by
EDWARD TOPHAM
1799

As for the stone itself, it is once again on display in the Science Museum of London.

chapter seven

# The Third Reptile

## *'The most perfect specimen known'*

Although it is not the largest specimen, the pride and joy of Whitby Museum is its fossil crocodile, *Teleosaurus chapmani* (WM770S). The museum itself is a magical collection of objects, brought together by Whitby people, and in particular by members of the Literary and Philosophical Society (which is still thriving), over the last 175 years. Its inclusiveness and peculiar eclecticism are a reflection of the town's history as a seaport and as a training ground for sailors. Spending an afternoon in the museum is like rummaging through the attic of a well-travelled, curious, inquisitive, obsessive, wealthy, slightly dotty and long-lived ancestor. The collection of fossils, particularly the reptiles, almost, but not quite, spoils this effect by being of international scientific importance. The great fossil crocodile was discovered in Whitby in 1824. George Young begins its story:

*"Account of a Fossil Crocodile recently discovered in the Alum-shale near Whitby*

> In the month of December 1824, an interesting discovery was made at Whitby. Brown Marshall, a well-known collector of petrifactions, observd, in the face of a steep cliff, not far from the town, part of the head of a large animal, standing out from the surface of the alum-shale, several yards above high water mark. Having, with no small labour and danger, succeeded in obtaining the head, he submitted it to my inspection; and I found it to correspond with some fossil heads

found here within these few years, which were considered as belonging to the Plesiosaurus. Being very desirous to procure a complete specimen of that animal, I directed him to be particularly careful in taking out the bones of the trunk, and especially the fin bones. After several days labour, attended with considerable peril, as the spot could not be reached but by the aid of ropes suspended from the upper part of the cliff, the whole specimen was got out. When the pieces into which it had parted were put together, and laid in the order in which they were found in the rock, I had the satisfaction of examining the specimen minutely; but what was my surprise, when, instead of an animal with *fins* for swimming, I found one with *legs* for walking; instead of a *Plesiosaurus*, I saw a *Crocodile*!

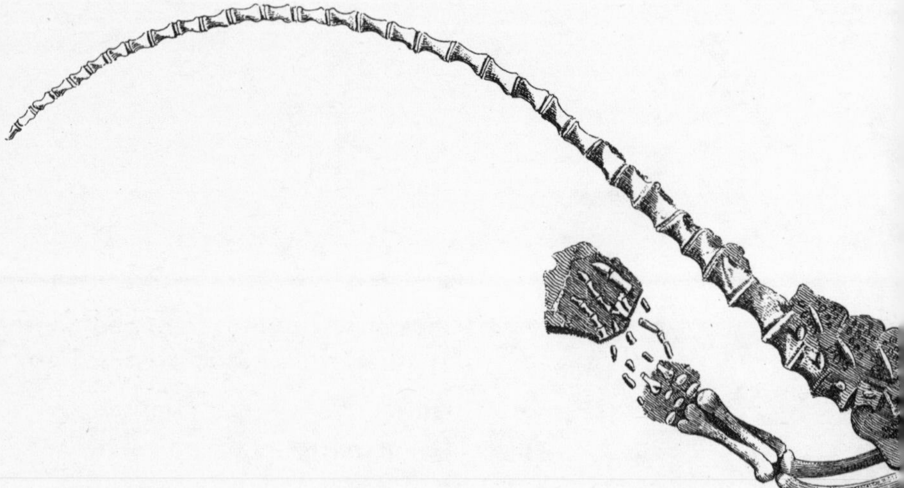

*The engraving of* Teleosaurus chapmani *which accompanied its first description in the* Edinburgh Philosophical Journal *in 1825.*

Most of the bones of both the hind-legs, with fragments of those of the forelegs were distinctly perceived. At the same time, the appearance of portions of the scaly crust of the animal, arranged in squarish compartments, as in the crocodile, made it easy to determine to what family the animal had belonged. This valuable relic of a former world was immediately purchased for the Whitby Literary and Philosophical Society, and conveyed to the Museum; and when some pains had been taken in removing a coating of alum-shale that had adhered to several parts of the mass, it became still more interesting."

George Young, 1825

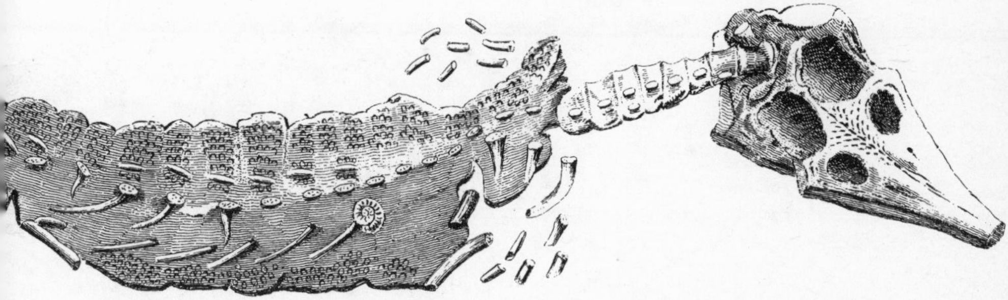

> "The most interesting article purchased this year is the FOSSIL CROCODILE, lately discovered by Brown Marshall, in the cliff near Saltwick. The existence of the Crocodile among the large animals imbedded in the alum-shale, had not been hitherto satisfactorily ascertained: but this specimen establishes the fact beyond all doubt; the animal being fully identified by the bones of its legs and feet, with some of the claws, and large portions of its scaly crust. From specimens of *amphibia* and large marine animals, now in the Museum, it appears, that at least four or five kinds of these bulky inhabitants of a former world have been lodged in our alum-shale: and the labours of the Society, in future years, may be the means of throwing much light on the nature and structure of these animals."
>
> Whitby Literary and Philosophical Society, 2nd annual report, 1825

The fossil was purchased for £7. Within two years the committee was becoming concerned about their prize specimen:

> "It is high time that our inestimate fossil, the great crocodile, and other organic remains, which lay exposed in the former Museum, should be put in cases adapted for their preservation; that they may continue to excite the wonder, and gratify the curiosity, of future ages."
>
> 4th annual report, 1827

When the new museum was built in the 1930s, the display of the large fossil reptiles caused some difficulties, as explained by the curator:

> "Chance again played an important part in the setting up of the Museum's most valuable possession, the fossil crocodile, named after Captain William Chapman. This had been placed in a flat table case with four stout legs. Before the plate glass was put on to protect this rare fossil, a camera was being supported, between two planks, with their ends rested on two pairs of steps. But some of the philosophers were so afraid that the photographer might slip, and fall on the top of the priceless object, that they suggested propping the table with its precious burden in an upright position; so that the photograph might be taken from the floor instead of from above.
>
> While the crocodile was in this position, Mr. L. T. Crawshaw, who had given much of his valuable time and assistance to arranging the other Saurians on the walls, saw the specimen in its upright position and noticed at once how much better it was so, than laid flat; so it was decided to have the table legs cut off and the case left as it was."
>
> F. M. Sutcliffe, 1946

The problems of display were temporarily solved, but the conservation of the great fossils became a problem – they began to rot. Several large plesiosaur and ichthyosaur specimens were built into the walls of the new Whitby Museum when it was erected in the 1930s. Over the next few decades rising damp began to penetrate the walls and to affect the precious fossil reptile collection. The pyrite which is necessary to the manufacture of alum – and was thereby the reason why these great skeletons first came to be accidentally uncovered – is also the source of their potential decay. As in the alum-making process itself, once the pyrite comes into contact with water and oxygen it begins to form sulphuric acid, which breaks down the aluminosilicates of the shale into powdery aluminium sulphate. The fossils simply rot away.

Building work was carried out at Whitby to stop and reverse the rising damp. This was successful, but unfortunately resulted in cracking in several specimens. Pyrite decay is caused by water vapour in the atmosphere. All museums with specimens from the Yorkshire alum shales try to control their humidity, or at least continually monitor its effects on their specimens.

Another problem arose in the last few decades. If you had gone into any of a dozen museums in Britain as recently as ten years ago, the fossil reptiles would have presented an awesome, but peculiar sight. At some time in the last 150 years (no one knows exactly when, though some of it dates from the 1920s), museum curators coated their specimens with various kinds of 'black consolidant'. Most often this resembled an oil-based varnish, which has darkened and become opaque with the passing years. In contrast, the matrix (the rock in which the fossil is embedded) was painted white. Again, no one knows quite why this was done. It may have been intended to enhance or dramatise the appearance of the fossils or a misguided attempt at helping their preservation. To modern eyes, with our renewed desire for what is 'natural', these tarted-up specimens were an increasingly unfortunate sight. A clean-up operation was begun.

In 1995 Whitby Museum received a grant from the Heritage Lottery Fund for conservation work on its fossil collection, and the first phase was completed by May 1996. Kate Andrew and her team returned in May 1997 to work on the museum's most precious specimen – the wonderful crocodile skeleton, *Teleosaurus chapmani*. As well as cleaning the specimen, the conservators found that a number of tail bones were actually placed in the wrong order.

The conservators first attempted to remove the black coating with hot and cold water, dilute ammonia solution, acetone, methylated spirit, propanol, orange oil-based solvents and solvent gel – none of which had any effect.

Next they turned to Nitromors paint remover and Ronstrip paint-removing poultice, which did the trick. The poultice was smeared on to the surface of the specimen, left for fifteen minutes, and then lifted off with a spatula. The fossil was then scrubbed with a toothbrush and cleaned with a mixture of de-ionised water and alcohol.

The fossil, which was named after the discoverer of the 1758 crocodile, is now cleaned and back on display in Whitby Museum. It awaits your visit.

chapter eight

# Whitby to the World

## *Statements concerning Captain James Cook*

### TERRA AUSTRALIS

> "Similarly the whole Indian Sea, along with the gulfs that are connected with it, the Arabian, Persian, Gangetic and the one that is properly called the Great Gulf, is entirely surrounded by land."
>
> Claudius Ptolemy, c. AD 140

Cook was not a discoverer. He went to look for a great continent and found only tiny islands in a vast ocean. If science is the continual dissolution or transformation of myth, then Cook acted like a scientist. It was Cook who banished for ever the idea of the Great Southern Continent.

Ptolemy's idea of the Indian Ocean, surrounded on all sides by land, including to the south, endured in Europe for a millennium and a half. The land that lay to the south of the Indian Ocean was unknown – *Terra Incognita Australis.* But there is no doubt that it existed, at least in the minds of all medieval and early Renaissance thinkers. Until the fifteenth Christian century the Great Southern Continent linked Africa, India and China. With the discovery of the new world it extended further west to join South America. It was the repository of strange animals (unicorns lived there) and was simultaneously seductive and fearsome. The voyages to the South Seas, from Bartholomeu Diaz in 1487 to James Cook in 1768, pushed this continent further south, though until Cook it remained a vast and fabulous notion.

In the late fifteenth century Portuguese adventurers, living on the edge of the world, began (for reasons we can only pretend to explain) to want to know more. Bartolomeu Diaz travelled south from Portugal, beyond the limits of the 'known world', sailing always within sight of the west coast of the continent of Africa. Each time he went a little further – and returned home. On one voyage, in 1487, he reached the end of Africa, and was blown eastwards around its southernmost tip by a storm. His crew became frightened and rebellious and he turned for home without exploring further. He had disconnected the Southern Continent from Africa.

Vasco da Gama followed the route of Diaz and reached Calicut in India in 1498. Ferdinand Magellan, the third great Portuguese maritime explorer, turned south-west and sailed around the southern tip of South America in 1519. His were the first European ships to enter the Pacific

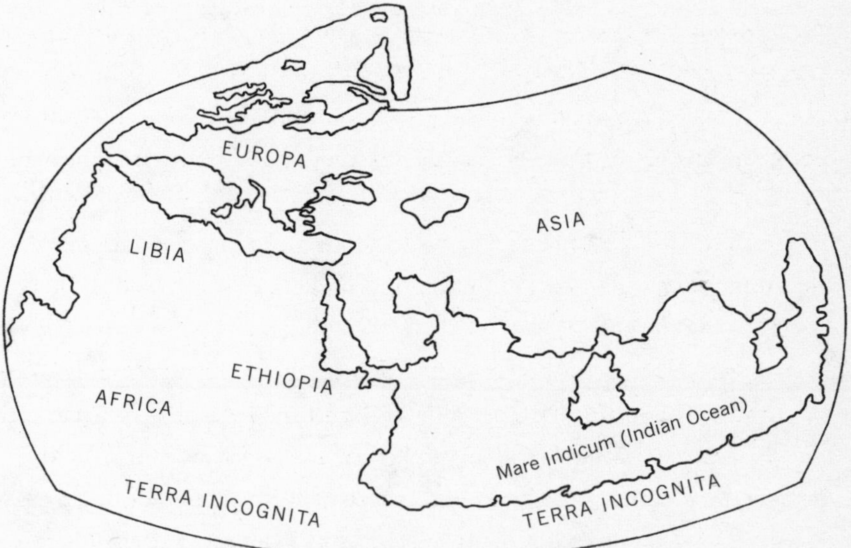


*Outline of the geography of the world, as described by Ptolemy in the first century* AD. *This is taken from a map published in Germany in 1486.*

Ocean. He sailed west for three months without sight of land, other than two tiny islets. His crew ate the leather off the ship's yards, they ate rats and sawdust. They reached the Ladrones and then went on to the Philippines, where Magellan offered to help put down a local rebellion. He bet on the wrong side and was killed by the rebels. His second-in-command, Sebastiano del Cano, sailed the only surviving ship from the flotilla, the *Victoria,* back to Europe, completing the first circumnavigation of the world. One hundred and seventy men died and four ships were lost on the voyage. Magellan separated the Southern Continent from the Americas and from Asia.

Abel Tasman in 1644 sailed around the south of Australia though, crucially, west and north of New Zealand. Tasman thereby disconnected Australia from the Southern Continent.

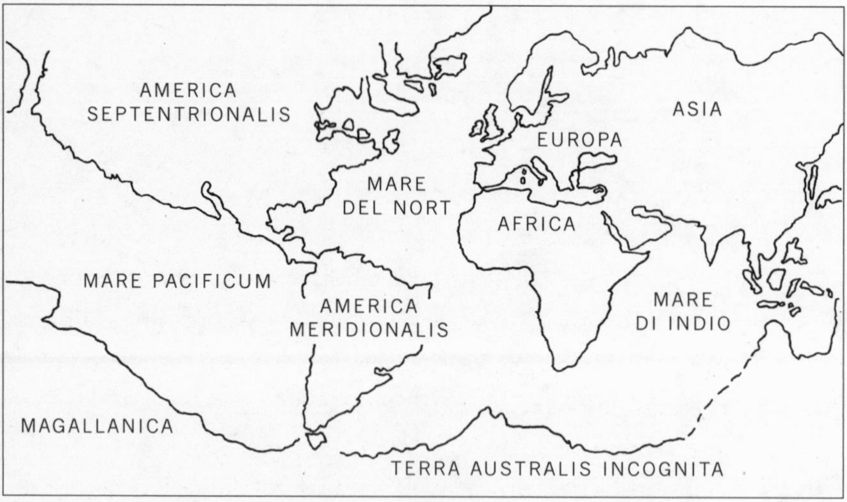


*Outline of a map of the world published in Holland, probably in the sixteenth century. Magellan had sailed around the tip of South America into the Pacific, but the width of the channel between the Americas and the Southern Continent was uncertain.*

Despite or because of Magellan, the Pacific was known only to be a vast area. There was still easily enough room in its southern reaches for a huge continent. Tantalising glimpses of land – New Zealand, South Georgia, the South Shetland Islands – might be its northern shore. Cook visited them all, and found them to be islands.

The Great Southern Continent existed because Ptolemy said it existed, and men were not of a mind to disbelieve the authorities of the classical past. Even as Ptolemy's scheme was shown to be wrong, it was simply amended and not scrapped. Was this typical of medieval thinking with its overplaced faith in authority? Or was it akin to the modern pragmatic rationalist, proposing only such changes as could be supported by physical evidence? Ptolemy's world was an artefact of rational imagination, it could not be overturned by logic. In the matter of geography, those who have gone and seen have won out over those who have sat and theorised.

## THE QUESTION OF SCIENCE

"He [Cook] was not original. He discovered nothing, in the modern scientific sense, he was no Priestley or Cavendish or Laplace; he invented nothing, he could not, like Harrison, make a chronometer. The genius of the matter-of-fact was the genius of the practical application of science, even at one remove; his exactitude was the fruit of his insistence on having, and his persistence in using, the best scientific instruments in a great age of scientific instrument making. He was fortunate, with his capacities, that he lived in that age. He was fortunate that he was no longer dependent

> exclusively on Halley's three L's, Lead, Latitude and Look-out; fortunate in Maskelyne's development of lunar distances, in the work of Harrison, Kendall, Bird, Dollond, Ramsden. He was fortunate in being confronted with what were, for him, the right problems at the right time. The time was fortunate that he was there to solve them."
>
> J. C. Beaglehole, 1974

Cook was a scientist because most scientists are involved in the practical work of measurement and recording.

Cook was not a scientist because he showed no abstract curiosity. Instead of questions, he brought answers.

> "Cook, the Whitby apprentice sailor, was a link between the march of 18th century scientific progress and the emergence of Britain as a world empire."
>
> *Whitby Gazette*, 1997

Joseph Banks, the young wealthy botanist who travelled on the *Endeavour*, brought back 30,000 specimens of plants and animals. He was knighted and became President of the Royal Society, a position he held for forty-two years. Banks was a great scientist.

Cook has been called the first scientific navigator. He is difficult to place between science and technology; it is easiest to think of him as a supreme technologist. In retrospect the technological tools – lunar tables for calculating longitude, improved telescopes and sextants, reliable chronometers, dietary methods for avoiding scurvy and thereby allowing long sea voyages – were there for anyone to use. Is the determined application of new technology an innovation in its own right? If Cook was not a conceptual scientist he was at least a technological innovator.

Geology is the study of the Earth. It was only when the entire Earth, or at least its configuration of land and sea, was 'known', that the science of geology became a possibility.

## THREE JOURNEYS

"So you are determined, Odysseus, my noble and resourceful lord, to leave at once for home and for your beloved Ithaca? Well, even so I wish you happiness. Yet had you any inkling of the full measure of misery you are bound to endure before you reach your motherland, you would not move from where you are, but you would stay and share this home with me, and take on immortality, however much you long to see that wife of yours . . .

I long to reach my home and see the happy day of my return. It is my never-failing wish. And what if the powers above do wreck me out on the wine-dark sea? I have a heart that is inured to suffering and I shall steel it to endure that too. For in my day I have had many bitter and shattering experiences in war and on the stormy sea. So let this new disaster come. It only makes one more."

Homer, c. 700 BC

Roberto Calasso tells us that Odysseus is the last of the heroes and the first of something else. He closes one world and opens another – the world we live in now. He is a hero in his intimacy with the gods. But he is a modern Everyman in his conniving spirit, in his postponement of gratification, in his determination to look for advantage rather than glory.

But what is this first modern man's odyssey? He is not in the age of parable or allegory. But the journey of Odysseus is a journey away from journeys. His travels take place on the sea, so he must find those who know nothing of it. A hero dressed as a servant, a sailor far from the sea, a master of verbal ingenuity with no opportunity to speak – he dies in a welter of self-negation.

Cook was and is a hero. Two and a half thousand years after the end of the age of heroes, Cook is not a child of Achilles, and perhaps not of Odysseus. Every book that is written about Cook lists those properties of his character which are to be seen as virtuous and heroic. The reader is aware of Cook's heroism, and is then told the ways in which he qualifies for this description. We thereby learn in what heroism resides. Heroism does not reside (any more) in the bluster of Achilles, nor in his anger, nor in his ritual humiliation of his enemy, though perhaps in his love of Patroclus. Nor does it quite reside in the cunning of Odysseus, nor in the despair which continually threatens to overwhelm him, but perhaps in his boldness, in his self-confidence, in his concern to get himself out of difficulties rather than to win glory.

Cook's most important quality is his deliberate erasing of his own personality. Almost everything that we know about him is connected to his ability to complete the tasks he has been set. He wants to tell us nothing about himself. Heroism is the undertaking of extraordinary tasks by an apparently ordinary person, but mostly by a person who has no personal interest in their own part in their own achievement. To be a hero you must not want to be a hero. More than that, the truth of what you achieve exists in relation to your own self-denial.

It has been said that the scientific revolution was an extended argument over different forms of truth. As Papal infallibility failed in the Reformation, and the Divine Right of Kings fell with the head of Charles I, as fixed social hierarchies crumbled, as wealth could, for the first time, be made as well as inherited, as the vernacular Bible was open to interpretation, where was authority to be located and who should decide both what was credible and who had access to the formation and holding of knowledge?

> "The more a body of knowledge is understood to be objective and disinterested, the more valuable it is as a tool in moral and political action. Conversely, the capacity of a body of knowledge to make valuable contributions to moral and political problems flows from an understanding that it was not produced and evaluated to further particular human interests."
>
> Steven Shapin, 1996

Cook meticulously followed the route set by Boyle, Newton and Halley, whose reports and speculations were peppered with insistences that they had nothing personally to gain from the world's belief in their work. Cook's reward for his modesty was acceptance into the community of knowledge-makers. On his return from his second voyage in 1775 he was made a Fellow of the Royal Society and awarded their highest honour, the Copley Medal for scientific achievement.

Scientific discoveries are better termed 'realisations'. If Cook lifted his head, as he surely must have, from his instruments and the conduct of his voyage, he would have been the first human being to see the geography of the world as we see it. Over his three epic voyages his charts meticulously recorded what was still to be recorded of the world, in a form that is directly understandable to us. We are the inheritors of Cook's vision. Cook was the first to hold a modern map of the world in his mind. The first modern geographer.

Cook's work meant that the configuration of the Earth could be said to be known. The boundaries were now described, there were no continents still to be discovered. The solid Earth could be held in men's minds as an entity. The Earth had become a suitable subject for study.

Once the map of the world was known, a new scheme for representing it was devised. The Earth was divided into a sea hemisphere and a land hemisphere. The result of this was, lo and behold, that western Europe sat precisely in the centre of the land hemisphere. This concoction neatly justified the mission of the Europeans to civilise, subdue and elevate the other peoples of the world.

> "Captain Cook came out of a labourer's cottage to take his place at the head of the masters of maritime exploration who worked at the great geographical problem of the Pacific. *Endeavour* was the name of the ship which carried him on his first voyage, and it was also the watchword of his professional life."
>
> Joseph Conrad

"Now when I was a little chap I had a passion for maps. I would look for hours at South America, or Africa, or Australia, and lose myself in all the glories of exploration. At that time there were many blank spaces on the earth, and when I saw one that looked particularly inviting on a map (but they all look like that) I would put my finger on it and say, When I grow up I will go there. The North Pole was one of these places, I remember. Well, I haven't been there yet, and shall not try now. The glamour's off. Other places were scattered about the Equator, and in every sort of latitude over the two hemispheres. I have been in some of them, and . . . well, we won't talk about that. But there was one yet – the biggest, the most blank, so to speak – that I had a hankering after.

True, by this time it was not a blank space any more. It had got filled since my boyhood with rivers and lakes and names. It had ceased to be a blank space of delightful mystery – a white patch for a boy to dream gloriously over. It had become a place of darkness. But there was in it one river especially, a mighty big river, that you could see on the map, resembling an immense snake uncoiled, with its head in the sea, its body at rest curving afar over a vast country, and its tail lost in the depths of the land. And as I looked at the map of it in a shop-window, it fascinated me as a snake would a bird – a silly little bird. Then I remembered there was a big concern, a Company for trade on that river. Dash it all! I thought to myself, they can't trade without using craft on that lot of fresh water – steamboats! Why shouldn't I try to get charge of one? I went along Fleet Street,

> but could not shake off the idea. The snake had charmed me."
>
> Joseph Conrad, *Heart of Darkness*, 1902

Perhaps if we really understood why Marlow went on his journey to see Kurtz, perhaps if we dared to see the banality of his impulse, we could put his dreams to one side. But what will boys dream of when the whole world is explored?

## WHITBY

After working in a shop in the fishing village of Staithes for two years, James Cook was apprenticed to John Walker, shipowner and trader in Whitby, in 1746 at the age of 18. During his apprenticeship he lived in his

*Whitby harbour in 1770, twenty years after Cook's apprenticeship in the port.*

master's house in Grape Lane in Whitby. He learned navigation while working on colliers taking coal down the the east coast of England from Newcastle to London. Whitby had no coal, but its natural harbour had been extended and the port enriched by the alum trade. By the 1740s it was a major centre for shipbuilding and coastal trading.

In an age when wars could still be won without destroying the vanquished and the victor, the Seven Years War ended with the Peace of Paris in 1763. England was, apart from the small matter of the American War of Independence, at peace with her neighbours and able to pursue other interests for the next three decades.

The *Endeavour* was built in Whitby by Thomas Fishburn's yard in 1764. She was squat and flat-bottomed, drawing only 14 feet, made for carrying coal on a coast with shallow shoals and harbours. She was bought by the navy for £2,840 10s. 11d. in 1768 for the purpose of Cook's voyage.

There has been a natural assumption that the choice of a Whitby-built collier must have been made by Cook, a Whitby-trained sailor who spent his early sailing life in these 'cats'. It was, though, a happy coincidence. Some anonymous officer on the Navy Board made the suggestion that a flat-bottomed collier would have the storage space required for a long journey, together with the manoeuvrability required for coastal navigation. Two possible ships, the *Valentine* and the *Earl of Pembroke*, were then lying in Shadwell docks in east London. The navy inspected both and recommended the latter, which was purchased from Thomas Milner, and renamed *His Majesty's Bark, the Endeavour* in 1768.

## EMBARKATION AND ORDERS

"Friday 26th August 1768. *Winds NBW, NW, WBS. Course S 21° E. Distce 23 Ml Latd in 49°30'. Longd in West from Greenwich 5°52' W. Bearings at Noon Lizard N21° W. Distance 23 miles.*

> First part fresh breeze and Clowdy, remainder little wind and Clear. At 2 pm got under sail and put to seas having on board 94 persons including Officers Seamen Gentlemen and their servants, near 18 months provisions, 10 Carriage guns 12 Swivels with good store of Ammunition and stores of all kinds."

James Cook, Journal of the *Endeavour*

Cook carried with him two sets of secret orders. The first concerned the observation of an astronomical phenomenon:

> "By the Commissioners for Executing the Office of Lord High Admiral of Great Britain &c
>
> *Secret*
>
> Whereas we have, in Obedience to the King's Commands, caused His Majesty's Bark the Endeavour, whereof you are Commander, to be fitted out in a proper manner for receiving such persons as the Royal Society should think fit to appoint to observe the Passage of the Planet Venus over the Disk of the Sun on the 3rd of June 1769, and for conveying them to such a Place to the Southward of the Equinoctial Line as should be judged proper for observing that Phaenomenon . . ."

This observation was part of an attempt to measure the distance of the Earth from the Sun. Cook's biographer explains the method.

> "The method for calculating the distance between the earth and the sun was the method of parallax: that is the method with which Cook, as a surveyor, was familiar, of observing angles with his theodolite at each end of his base line, and working out trigonometrically therefrom the distance to his marker. But now, though the base line might be something like the radius of the earth in length, the marker – the sun – was so far away that the parallax counted for hardly anything, and an intermediate help – a sort of observational stepping-stone – was needed. This intermediate help or point was provided by the planet Venus.
>
> It was provided – or we may speak in the present tense and say it is provided in human lives but rarely: at those times only when Venus is in a direct line between the earth and the sun, and its black shadow as this passes across the face of the sun can be observed and timed. The time taken by such a 'Transit of Venus' depends on the rate at which the line joining the observer's eye to Venus sweeps across the face of the sun. If the earth were not rotating, this line would move at the same speed for all observers, but because it does rotate, the observer's end of the line moves at a speed determined by his position on the earth and by the apparent size of the earth as seen from Venus. The different times taken for the transit, as measured by different observers, can with much calculation yield the

parallax, and hence the total distance from the earth to Venus and earth to sun."

J. C. Beaglehole, 1974

The secret instructions justify the claim that this was the first voyage in human history that was made purely in search of knowledge. But there was another 'Sealed Packet' of secret orders, referred to in the first set.

"When this service [the observations of the Transit of Venus at Tahiti] is perform'd you are to put to Sea without Loss of Time, and carry into execution the Additional Instructions contained in the inclosed Sealed Packet."

The contents of this Sealed Packet had been thought to be lost, but were rediscovered in naval records in the 1920s:

"Whereas the making Discoverys of Countries hitherto unknown and the Attaining a Knowledge of distant parts which though formerly discover'd have yet been but imperfectly explored, will redound greatly to the Honour of this Nation as a Maritime Power, as well as to the Dignity of the Crown of Great Britain, and may tend greatly to the advancement of the Trade and Navigation thereof; and whereas there is reason to imagine that a Continent or Land of great extent may be found to the Southward . . .

You are to proceed to the southward in order to make discovery of the Continent abovementioned until you arrive in the Latitude of 40°, unless you sooner fall

> in with it. But not having discover'd it or any signs of it in that Run, you are to proceed in search of it to the Westward between the Latitude before mentioned and the Latitude of 35° until you discover it, or fall in with the Eastern side of the Land discover'd by Tasman and now called New Zealand."

Cook was also given the following letter.

*"To The Flag Officers, Captains & Commanders of His Majesty's Ships and Vessels to whom this shall be exhibited.*
30th July 1768

> Whereas we have directed Lieut. James Cook to proceed in His Majesty's Bark the Endeavour upon a particular service, you are hereby required and directed not to demand of him a sight of the Instructions he has received from us for his proceedings on the said service, nor upon any pretence whatever to detain him, but on the contrary to give him any assistance he may stand in need of, towards enabling him to carry the said instructions into execution.
>
> E. HAWKE
> P. BRETT
> C. SPENCER"

Despite the theatricality of the Sealed Packet, it seems certain that Cook and his 'passengers,' if not his officers and crew, were aware of the real purpose of their mission – to discover and map the Great Southern Continent, *Terra Incognita Australis,* with the possibility of claiming it for the British Crown.

## DEALINGS WITH THE NATIVES

"Hints offered to the consideration of Lieutenant Cooke, 10th August 1768

> To exercise the utmost patience and forebearance with respect to the Natives of the several Lands where the Ship may touch. To check the petulance of the Sailors, and restrain the wanton use of Fire Arms. To have it still in view that sheding the blood of those people is a crime of the highest nature . . . They are the natural, and in the strictest sense of the word, the legal possessors of the several Regions they inhabit . . . They may naturally and justly attempt to repell intruders, whom they may apprehend are come to disturb them in the quiet possession of their country, whether that apprehension be well or ill-founded . . . The Natives when brought under should be treated with distinguished humanity, and made sensible that the Crew still considers them as Lords of the Country."

Letter from Lord Morton, President of the Royal Society, to James Cook, prior to his journey on the *Endeavour*

"You are to endeavour by all proper means to cultivate a friendship with the Natives, presenting them such trifles as may be acceptable to them, exchanging with them for Provisions . . . and showing every kind of civility and regard. But as Capn Wallis has represented the Island to be very populous, and the Natives to be rather treacherous than otherwise you are to be

Cautious not to let your self be surprized by them, but to be constantly on your guard against any accident."

Extract from Cook's official orders

Notice from Cook to every member of his crew:

"RULES to be observ'd by every person in or belonging to his Majestys Bark the Endeavour, for the better establishing a regular and uniform Trade for Provisions &c with the Inhabitants of Georges Island [Tahiti].

1st To endeavour by every fair means to cultivate a friendship with the Natives and to treat them with all imaginable humanity.

2d A proper person or persons will be appointed to trade with the Natives for all manner of Provisions, Fruit, and other productions of the earth; and no officer or Seaman, or other person belonging to the Ship, excepting such as are so appointed, shall Trade or offer to Trade for any sort of Provisions, Fruit, or other productions of the earth unless they have my leave to do so.

3d Every person employ'd a Shore on any duty what soever is strictly to attend to the same, and if by neglect he looseth any of his Arms or working tools, or suffers them to be stole, the full Value thereof will be charge'd againest his pay according to the Custom of the Navy in such cases, and he shall receive such further punishment as the nature of the offence may deserve.

4th The same penalty will be inflicted on every person who is found to imbezzle, trade or offer to trade with any part of the Ships Stores of what nature soever.

5th No sort of Iron, or any thing that is made of Iron, or any sort of Cloth or other usefull or necessary articles are to be given in exchange for any thing but provisions."

"October 1769

. . . I rowed round the head of the Bay but could find no place where I could land on account of the great surf which beat every where upon the shore/ this made me resolve upon takeing one of 2 boats I saw coming in to the Bay, and it so happend that one of them came so near us we calld to them to come to us, but instead of that they endeavoured to get away, upon which I ordered a musquet to be fired over their heads, thinking that it would either oblige them to surrender or jump over board as we was at this time but a few yards from them/ but in this I was mistaken/ for they took to their arms or what ever they had in the Boat and begun to attack us, this caused us to fire upon them and it unfortunatly happend that 2 or 3 was killd and one wounded and 3 jumped overboard unhurt/ these we took up and brought on board the Ship where they had cloase and victuals given them and where treated in the best manner . . . the oldest might be about 18 or 19 and the youngest about 9 or 10. I can by no means justify my conduct in attacking and killing the people in this boat who had given me no just provication and was wholy igernorant of my design and had I had the least thought of their making any resistance I would not so much as have looked at them but when we was once a long side of them we must either had stud to be knockd on the head or else retire and let them gone off in triumph and this last they would of Course have attributed to their own bravery and our timourousness. Thus ended the most disagreable day My life has yet seen, black be the mark for it and heaven send that such may never return to embitter future reflection."

James Cook, Journal of the *Endeavour*

## DEALING WITH THE CREW

"Wednesday 30th November 1768

Punished Robt Anderson Seaman and Willm Judge Marine with twelve lashes each, the former for leaving his duty a Shore and attempting to disert from the Ship, and the latter for useing abusive language to the Officer of the Watch, and John Readon Boatswains mate with twelve lashes for not doing his duty in punishing the above two Men."

James Cook, Journal of the *Endeavour*

"Wednesday 23rd [May 1770, Off the Barrier Reef]

Last night some time in the Middle watch, a very extraordinary affair happend to Mr Orton my Clerk, he having been drinking in the Evening, some Malicious person or persons in the Ship took the advantage of his being drunk and cut off all the cloaths from off his back, not being satisfied with this they some time after went into his Cabbin and cut off a part of both his Ears as he lay asleep in his bed. The person whome he suspected to have done this was Mr Magra one of the Midshipmen, but this did not appear to me upon inquirey. However as I know'd Magra had once or twice before this in their drunken frolicks cut of his Cloaths and had been heard to say (as I was told) that if it was not for the Law he would Murder him, these things consider'd induce'd me to think that Magra was not altogether innocent. I therefore, for the present dismiss'd him the quarter deck and susspended him from doing any duty in the Ship, he being one of those gentlemen, frequently found on board Kings Ships, that can very well be spared, or to speake more planer good for nothing. Besides it was necessary in me to show my immediate resentment againest the person on whome the suspicion fell least they should not have stoped here. With respect to Mr Orton he is not a man without faults, yet from all the enquiry I could make, it evidently appear'd to me that so far from deserving such treatment he had not designedly injured any person in the Ship, so that I do and shall all ways look upon him as an enjure'd man.

Some reasons might however be given why this misfortune cam upon him in which he himself was in some measure to blame, but as this is only conjector and would tend to fix it upon some people in the Ship whome I would fain believe would hardly be guilty of such action, I shall say nothing about it unless I hereafter discover the Offenders which I shall take every method in my power to do, for I look upon such proceedings as highly dangerous in such Voyages as this and the greatest insult that could be offer'd to my authority in this Ship, as I have always been ready to hear and redress every complaint that have been made against any Person in the Ship."

James Cook, Journal of the *Endeavour*

## LOTUS EATING

"We disembarked to draw water, and my crews quickly set on to their midday meal by the ships. But as soon as we had a mouthful and a drink, I sent some of my followers inland to find out what sort of human beings might be there, detailing two men for the duty with a third as messenger. Off they went, and it was not long before they were in touch with the Lotus-eaters. Now it never entered the heads of these natives to kill my friends; what they did was to give them some lotus to taste, and as soon as each had eaten the honeyed fruit of the plant, all thoughts of reporting to us or escaping were banished from [their minds]. All they now wished for was to stay where they were with the Lotus-eaters, to browse on the lotus, and to forget that they had a home to return to. I had to use force to bring them back to the ships, and they wept on the way, but once on board I dragged them under the benches and left them in irons. I then commanded the rest of my loyal band to embark with all speed on their fast ships, for fear that others of them might eat the lotus and think no more of home. They came on board at once, went to the benches, sat down in their proper places, and struck the white surf with their oars."

Homer, c.700BC

"Sunday 9th [July 1769, King George's Island (Tahiti)]

When sometime in the Middle Watch Clement Webb & Saml Gibson both Marines & young Men found means to get away from the Fort (which was no hard matter to do) & in the morning were not to be found, as it was known to every body that all hands were to go on board on the monday morning & that the ship would sail in a day or 2, there was reason to think that these men intended to stay behind, However I was willing to wait one day to see if they would return before I took any steps to find them.

Monday 10th

The 2 Marines not returning this morning I began to enquire after them & was inform'd by some of the Natives that they were gone to the Mountains & that they had got each of them a Wife & would not return, but at the same time no one would give us any Certain intelligence where they were, upon which a resolution was taken to seize upon as many of the Chiefs as we could, this was thought to be the readiest way to induce the other natives to produce the two men.

About 9 o'clock in the evening Web the Marine was brought in by some of the Natives and sent on board, he inform'd me that the Petty officer & the Corporal that had been sent in quest of them were disarm'd and seiz'd upon by the Natives and that Gibson was with them. Immidiatly upon getting this information I dispatch'd Mr Hicks away in the Long boat with a Strong party of men to resque them . . .

The guides conducted Mr Hicks to the place before daylight and he recover'd the men without the least opposission and return'd with them about 7 oClock in the Morning of

Tuesday 11th

. . . Thus we are likly to leave these people in disgust with our behaviour towards them, owing wholy to the folly of two of our own people for it doth not appear that the natives had any hand in inticeing them away and therefore were not the first aggressors, however it is very certain that had we not taken this step we never should have recover'd them.

When I cam to examine these two men touching the reasons that induc'd them to go away, it appear'd that an acquentence they had contracted with two Girls and to whome they had stron[g]ly attache'd themselves was the sole reason of their attempting to stay behind."

James Cook, Journal of the *Endeavour*

## HEALTH, SICKNESS AND DEATH

Cook's first voyage was not simply a journey that had not been tried before, it was a probable sentence of death for half his crew. In the end he returned after two years and eleven months with fifty out of ninety-four crew and passengers alive and tolerably well (extra crew had been taken on board at Batavia). Most ships lost as many men on a one-year journey to the South Seas, from scurvy or dysentery.

"Thursday 13 [April 1769, just arrived and anchored off Tahiti]

At this time we had but a very few men upon the Sick list and thes had but slite complaints, the Ships compney had in general been very healthy owing in a great measure to the Sour krout, Portable Soup and Malt; the two first were serve'd to the People, the one on Beef Days and the other on Banyan Days [naval term for meatless days, from *Banian*, a Hindu, and therefore vegetarian, trader], Wort was made of the Malt and at the discrition of the Surgeon given to every man that had the least symptons of Scurvy upon him, by this Means and the care and Vigilance of Mr Munkhous the Surgeon this disease was prevented in getting a footing in the Ship. The Sour Krout the Men at first would not eate untill I put in practice a Method I never once knew to fail with seamen, and this was to have some of it dress'd every Day for the Cabbin Table and permitted all the Officers without exception to make use of it and left it to the option of the Men either to take as much as they pleased or none at all; but this practice was not continued above a week before I found it necessary to put every one on board to an Allowance, for such are the Tempers and dispossions of Seamen in general that whatever you give them out of the Common way, altho it be ever so much for their good yet it will not go down with them and you will hear nothing but murmurings gainset the man that first invented it; but the Moment they see their Superiors set a Value upon it, it becomes the finest stuff in the World and the inventer an honest fellow."

James Cook, Journal of the *Endeavour*

Cook put in at the island of Java to collect a supply of fresh water. He was extremely diligent about the source of his supply, but in this case the water was infected. For the month-long voyage from Java to the Cape of Good Hope the *Endeavour* was a ship full of sickness and death.

"Thursday 24th [January 1771]

In the AM died Jno Truslove Corpl of marines, a Man much esteem'd by every one on board. Many of our people at this time lay dangerously ill of Fevers and fluxes. We are inclinable to attribute this to the water we took in at Princes Island and have put lime into the Casks to purifie it.

Friday 25th.

Light Airs and Calms, hot and Sultry weather. Departed this life Mr Sporing a Gentleman belonging to Mr Banks's retinue.

Saturday 26th.

First part little wind, the remainder Calm and very hot.

Set up the Topmasts rigging and clean'd Ship between decks and wash'd with vinegar.

Sunday 27th.

Little wind and some times calm. Departd this Life Mr Sidney Parkinson, Natural History Painter to Mr Banks, and soon after Jno Ravenhill, Sailmaker, a Man much advanced in years.

Tuesday 29th.

Very variable weather, some times squally with rain, other times Little wind and Calms. In the night Died Mr Charls Green who was sent out by the Royal Society to Observe the Transit of Venus; he had long been in a bad state of hilth, which he took no care to repair but on the contrary lived in such a manner as greatly promoted the disorders he had long upon him, this brought on the Flux which put a period to his life.

Wednesday 30th.

First and Latter parts Moderate breezes and clowdy weather, the middle Squally with rain Thunder and Lightning. Died of the Flux Sam Moody and Francis Hate, two of the Carpenters Crew.

Thursday 31st.

First part moderate and fair, the remainder frequent squalls attended with showers of rain. In the Course of this 24 hours we have had four Men died of the Flux, viz. Jno Thompson Ships Cook, Benj Jordan Carpenters Mate, James Nicholson and Archd Wolfe

Seamen. A Melancholy proff of the Calamitous Situation we are at present in, having hardly well men enough to tend the Sails and look after the Sick, many of the latter are so ill that we have not the least hopes of their recovery.

February 1771. Friday 1st.

Fresh gales with flying showers of Rain. Clean'd between Decks and Wash'd with Vinegr.

Saturday 2nd.

A Fresh Trade and mostly fair weather. Departed this Life Danl Roberts Gunners Servant who died of the flux. Since we have had a fresh Trade wind this fatall disorder hath seem'd to be at a stand, yet there are several people which are so far gone and brought so very low by it that we have not the least hopes of their recovery.

Sunday 3rd.

Ditto weather. Departed this Life John Thurman Sailmakers assistant.

Monday 4th.

A Fresh Trade wind and hazey weather, with some Squalls attend with small rain. Unbent the Main Topsail to Repair and bent a nother. In the night died of the Flux Mr John Bootie Midshipman and Mr Jno Gathrey Boatswain.

Wednesday 6th.

A Fresh Trade wind and fair weather. In the night died Mr John Monkhouse Midshipman and Brother to the late Surgeon.

Tuesday 12th.

Gentle breezes and fair weather. At 7 in the AM died of the flux after a long and painfull illness Mr John Satterly, Carpenter, a Man much Esteem'd by me and every Gentleman on board, in his room I appoint George Knowel one of the Carpenters Crew, having only him and one More left.

Thursday 14th.

Moderate breezes and Clowdy with some Showers of rain. Departed this Life Alexr Lindsey Seaman; this man was one of those we got at Batavia and had been some time in India.

Friday 15th.

Weather as yesterday. Died of the flux Danl Preston Marine.

Thursday 21st.

First and Middle parts fair weather, Latter Squally attended with Showers of rain. In the night died of the flux Alexr Simpson a very good Seaman. In the Morning Punished Thos Rossiter with Twelve Lashes for geting Drunk, grossly Assaulting the Officer of the Watch and beating some of the Sick.

Sunday 24th.

In the AM took the oppertunity of a fine morning to Stay the Main Mast and set up the Topmast Rigging. Saw an Albatross.

Wednesday 27th.

Gales and clowdy. In the AM Died of the Flux Henry Jeffs, Emanuel Pharah and Peter Morgan Seamen, the last came sick on board at Batavia of which he never recoverd and the other two had long been past all hopes of recovery, so that the death of these three men in one day did not in the least alarm us; on the contrary we are in hopes that they will be the last that will fall a Sacrefice to this fatal desorder, for such as are now ill of it are in a fair way of recovering."

James Cook, Journal of the *Endeavour*

## NAVIGATING THE PACIFIC

Cook described the vessels of the Tahitians, and understood their abilities to travel across the Pacific:

"Their Proes or Canoes large and small are rowed and steer'd with paddles and notwithstanding the large ones appear to be very unwieldy they manage them very dextrusly and I believe perform long and distant Voyages in them, otherwise they would not have the knowlidge of the Islands in these seas they seem to have."

James Cook, Journal of the *Endeavour*

"The Tahitians who encountered Cook's men were also heirs of a culture of expert navigators. During the two millennia before 500AD in European reckoning, the vast area of sea and islands stretching between Hawaii, New Zealand and the western coasts of South America had already been crossed by Polynesians, a feat far beyond any navigational skill then possessed by the ancestors of Cook. Some of the 'experimental gentlemen' on board his ships meditated on the Polynesian achievement however little they knew of its true scope. [Forster wrote:] "This exercise of their memories, and of other mental faculties confirming by their own experience the truth of the phenomena communicated to them by their parents and teachers gives them as it were a predilection for the examination of the truth."

Simon Schaffer, 1996

The navigational directions for travelling from island to island across thousands of miles of featureless sea were passed on in 'stories' about the stars. These were myths by which the listener was brought to an understanding. Some have survived to remind us of the beauty and complexity of a woven oral fabric.

"[The sailor] came across an old woman sitting at the door of her house (the Pleiades) on whom he played some familiar trick, which caused her to run away westwards (i.e. decline towards her setting). Next he met a man coming in a canoe from the east (steered by way of Aldebaran in the constellation of Taurus).

With him he held converse, until the old woman, who had run away from him fell into the sea (the Pleiades set)."

Giorgio de Santillana, 1961

## THE DEMISE OF TERRA AUSTRALIS

Cook separated New Zealand from the Great Southern Continent on 10 March 1770. This is his journal entry for that day.

"Saturday 10th

At sunset the Southernmost point of land which I afterwards named *South Cape* and which lies on the Latde 47°19'S, Longd 192°12' West from Greenwich bore N 38° E distant 4 Leagues and the westernmost land in sight bore N 2° East, this last was a small isld lying off the point of the Main. I began now to think that this was the southernmost land and that we should now be able to get around it by the west, for we have had a large hollow swell from the SW ever since we had the last gale of wind from that quarter which makes me think that there is no land in that quarter."

Cook was right of course – there is no land south-west of New Zealand for 2,000 miles. He wrote his summary on his return to England in 1771 and, like a good scientist, made the case for further researches:

"But to return to our own Voyage which must be allow'd to have set a side the most if not all the arguments and proofs that have been advance'd by different Authors to prove that there must be a Southern Continent, I mean to the northward of 40° S for what may lay southward of that latitude I know not . . .

Thus I have given my Opinion freely and without prejudicy not with any view to discourage any future attempts being made twoards discovering the *Southern Continent*, on the Contrary, as this Voyage will evidently make it appear that there is left but a small space to the Northward of 40° where the grand Object can lay, I think it would be a great pitty that this thing which at times has been the object of many ages and Nations should not now be wholy clear'd up."

James Cook, Journal of the *Endeavour*

*Captain Cook as hero: stained-glass window at Marton church in Cleveland.*

Once the geography of the world was known, it could be considered as a whole. And this consideration could, in time, be fashioned into a science. Once the continents and seas of the world were known, there were no places left for dinosaurs or ichthyosaurs or plesiosaurs to live. Extinction turned from an argument into a scientific truth.

chapter nine

# Devils, Romans and Rocks

## *Geology and myth at Filey Brigg*

"Before geological investigations were much attended to in these parts, it was supposed that this celebrated reef was formed by some Typhoeus, or by some other supernatural being of a worse name, with the intention of continuing it to Denmark and Norway, so that the witches, who, braving the perils of the deep, had been accustomed to come over to this country in egg shells, might arrive by dry land; but fortunately this evil-disposed personage lost his hammer before his wishes were realized, so his labours ceased."

Martin Simpson, 1855

"Two dark marks on the shoulders of the haddock are, by the legends of many lands, attributed to the Evil One; though by another legend the fish is said to have been the one caught by St Peter, at his Divine Master's command, in the Sea of Galilee, and the marks to have been those of the apostle's finger and thumb, made in holding the fish while he extracted the piece of money for the tribute from its mouth.

The Yorkshire legend, however, brings the origin of the marks nearer home. According to it, the Devil was the builder of the well-known dangerous ridge of rocks known as Filey Brigg. As he was proceeding with this work, he, by chance, dropped his hammer into the water. Diving, in haste to recover it, he by mistake seized a haddock instead of the hammer. Since then the whole species has borne upon its sides the marks of the infernal hand, and shall bear them to the end of time."

Thomas Parkinson, 1888

"A tradition exists that the Spittal – a projecting spur of rocky material on the south side of the Brigg, and partially visible at very low tides, is a remnant of a Roman Harbour, but no warrant for this tradition has yet been discovered.

The name 'Filey Bridge' upon the old one-inch map is one among many absurd mistakes in an apparent effort to 'refine' the ruggedness of the placenames of this Northland. The word 'Brigg' is obviously not associated with the idea of a bridge, but is the Scandinavian word meaning a pier. This pier-like promontory is composed of the Lower Calcareous Grit (Middle Oolites), and it continues that strike of the rocks which we have traced from west of Scarborough. The platform is 200 to 500 feet wide with a gentle dip to the south and is broken by great steps, and holes. Opposing its escarpment to the flow of the tides down the coast, the north edge is eroded into many beautiful coves where – especially when the wind is from the north-east – the waves dash with tremendous breakers. When this scene is most picturesque, because most turbulent, the Brigg becomes very dangerous. Visitors have by exceptionally heavy waters been swept from the rocks and drowned. In suitable weather and at low tides one can walk far along the reef, and its pools and shallows, crowded with animal and vegetable life, are to the marine zoologist among the most delightful spots on the Coast."

Percy Fry Kendall and Herbert E. Wroot, 1924

"Filey Bay is terminated on the north side by the straight flank of a long promontory of hard Corallian rocks overlain for half its length by 100ft (30m) of boulder clay. The landward drift-covered part of this peninsula is known as Carr Naze; the sea-swept rocky part as the Brigg.

The Brigg is an extraordinary reef, an extension of Carr Naze with the boulder clay stripped off. It forms a mile-long natural and dangerously projecting reef covered at high tide and has at its extremity a buoy with a mysterious-sounding haunting clang. The sea has broken up the Brigg into a series of great steps and holes with the edges facing the northerly exposed side. Huge, rectangular, broken-off masses of rock lie along these edges in a rough imbricate pattern as though they had been turned over into the same position by huge seas."

Derek Brumhead, 1979

chapter ten

# The Fourth Reptile

## *The disappearing plesiosaur*

There were many more fossil reptiles dug from the alum shales of the Yorkshire coast than are known to exist today. Many of them were bought by private collectors, and have never been seen since. They may be stored in attics or basements, but more likely they have been broken up and discarded by unknowing descendants. It is, though, unusual for a major fossil specimen in a public collection simply to go missing – particularly without the fact of its disappearance being recorded. But that is what happened to Scarborough Museum's 'most valuable Fossil Animal' some time between 1837 and 1840.

The fossil first appears in print in the second edition of George Young's *A Geological Survey of the Yorkshire Coast*, published in 1828:

> "Along with the ichthyosaurus, our alum shale contains the remains of another large animal, called the PLESIOSAURUS. The noble specimen discovered at Lyme Regis, in 1823, has enabled us to identify several of the bones of this rare fossil animal; of which no entire specimen, and no head, has yet been found in our strata. A specimen showing a great part of the spine, several of the paddle bones, portions of ribs, and some of the humero-sternal bones, is the best that we have obtained. Other portions of the vertebral column, of the ribs, the fin bones, and the breast bones, are in the Museum; and from the size of part of them, it appears that some of these animals have been truly gigantic."

The next mention is in the Report of the Council of the Scarborough Philosophical Society, owners of Scarborough's Rotunda Museum, dated 31 August 1831:

> "Whenever any very rare or valuable relic has been found, belonging to the coast in the vicinity of Scarborough, the Council felt it a duty not to forego the opportunity of purchasing it, since the specimen might never afterwards be obtained, and thus lost from the neighbourhood altogether. In one instance, where the expense of the object exceeded what the condition of the funds would allow, the very spirited curator of Geology (Mr Stickney) became the instrument of furnishing the Society with the most valuable Fossil Animal in the collection, and perhaps the largest Plesiosaurus in Europe.
>
> (This Skeleton has since been minutely described to the members by Mr Dunn, and his memoir read to the Geological Society of London.)"

Mr Dunn's description is as follows:

"On a large species of Plesiosaurus in the Scarborough Museum.

> The animal was discovered by Mr. Marshall, of Whitby, imbedded in a hard rock belonging to the upper lias beds, situate between Scarborough and Whitby, near the place where that gentleman had formerly discovered the remains of a crocodile. The skull and cervical vertebrae are wanting, but the rest of the skeleton is pretty entire and

> measures from the anterior dorsal to the last coccygeal vertebra nine feet six inches. The entire animal, with the head and neck is estimated to have been nineteen feet long."

There follows a detailed description of the skeleton, before the report finishes:

> "The author has given as full an account of all the bones of this interesting animal as the hardness of the imbedding rock, or the safety of the specimen, would allow of their examination."

There is another mention in a survey of Lias reptiles by Williamson in 1837:

> "The *Plesiosaurus* varies in size, but from a comparison of detached bones with those of more complete specimens, the animal appears to have averaged from fifteen to twenty feet in length. The large specimen in the Scarborough museum would measure, if complete, at least eighteen feet. Its general characters resemble those of the *Plesiosaurus* common in the lias of Dorset."

From the description, this was a gigantic plesiosaur skeleton – with the head and neck missing – 'the most valuable Fossil Animal in the collection, and perhaps the largest Plesiosaurus in Europe'. It must have been the centrepiece of the museum. And yet Dunn's report is the last we ever hear of it. Richard Owen, the most celebrated fossil anatomist of the day, conducted a survey of the fossil reptiles of Britain on behalf of the British Association in 1839 and 1840. Every major specimen in the country is described in Owen's work. But, just two or three years after Williamson's mention, and just nine years or so after Dunn's detailed description, the Scarborough plesiosaur is missing from Owen's report. It was, so far as anyone knows, never seen again.

chapter eleven

# Geological Journeys

## *Order, harmony, morality and time*

### GEOLOGY AND ORDER

"If we suppose an intelligent traveler taking his departure from our metropolis, to make from that point several successive journeys to various parts of the island, for instance to South Wales, or to North Wales, or to Cumberland, or to Northumberland, he cannot fail to notice (if he pays any attention to the physical geography of the country through which he passes) that before he arrives at the districts in which coal is found, he will first pass a tract of clay and sand; then another of chalk; that he will next observe numerous quarries of the calcareous freestone employed in architecture; that he will afterwards pass a broad zone of red marly sand; and beyond this will find himself in the midst of coal mines and iron furnaces. This order he will find to be invariably the same, whichever of the routes above indicated he pursues; and if he proceeds further, he will perceive that near the limits of the coal-fields he will generally observe hills of the same kind of compact limestone, affording grey and dark marbles, and abounding in mines of lead and zinc; and at a yet greater distance, mountainous tracts in which roofing slate abounds, and the mines are yet more valuable; and lastly, he will often find, surrounded by these slaty tracts, central groups of granitic rocks. The intelligent enquirer, when he has once generalised these observations, can scarcely fail to conclude that such coincidences cannot be casual; but that they indicate a regular succession and order in the mineral masses constituting the Earth's surface; and he must at once perceive that, supposing such an order to exist, it must be of the highest importance to economical as well as scientific objects, to trace and ascertain it."

William Conybeare and William Phillips, 1822

## GEOLOGY AND HARMONY

"A person proceeds from London to North Wales. After passing low diluvial plains about London, he climbs, by a long slope, the chalk-hills of Oxfordshire and Berkshire; then crosses vales of clay and sandstone, ascends a range of oolitic limestone; traverses wide plains of blue and red marl; arrives in districts where coal, iron, and limestone abound; and finally sees Snowdon composed of slate. And if, in proceeding from London to the Cumberland lakes, he finds the same succession of low plains, chalk-hills, clay vales, oolitic limestone ranges, blue and red clays, coal, iron, and limestone tracts, succeeded by the slate rocks which compose the well-known summit of Skiddaw, will he not conclude that something beyond mere chance has brought together these rocks in such admirable harmony? Will he not have reason to conjecture, that, in the interior of the earth, regularity of arrangement must prevail."

John Phillips, 1829

## GEOLOGY AND MORAL CONDITION

"If a stranger, landing at the extremity of England, were to traverse the whole of Cornwall and the North of Devonshire; and crossing to St David's, should make the tour of all North Wales; and passing thence through Cumberland, by the Isle of Man, to the south-western shore of Scotland, should proceed either through the hilly region of the Border Counties, or, along the Grampians to the German Ocean; he would conclude that Britain was a thinly populated sterile region, whose principal inhabitants were miners and mountaineers.

Another foreigner, arriving on the coast of Devon, and crossing the Midland Counties, from the mouth of the Exe, to that of the Tyne, would find a continued succession of fertile hills and valleys, thickly overspread with towns and cities, and in many parts crowded with a manufacturing population, whose industry is maintained by the coal with which the strata of these districts are abundantly interspersed.

A third foreigner might travel from the coast of Dorset to the coast of Yorkshire, over elevated plains of oolitic limestone, or of chalk; without a single mountain, or mine, or coal-pit, or any important manufactory, and occupied by a population almost exclusively agricultural.

Let us suppose these three strangers to meet at the termination of their journeys, and to compare their respective observations; how different would be the results to which each would have arrived, respecting the actual condition of Great Britain. The first would represent it as a thinly peopled region of barren mountains; the second, as a land of rich pasture, crowded with a flourishing population of manufacturers; the third, as a great corn field, occupied by persons almost exclusively engaged in the pursuits of husbandry.

These dissimilar conditions of three great divisions of our country, result from differences in the geological structure of the districts through which our three travellers have been conducted. The first will have seen only those north-western portions of Britain, that are composed of rocks belonging to the primary and transition series: the second will have traversed those fertile portions of the new red sandstone formation which are made up of the detritus of more ancient rocks, and have beneath, and near them, inestimable treasures of mineral coal: the third will have confined his route to wolds of limestone, and downs of chalk, which are best adapted for sheep-walks, and the production of corn.

Hence it appears that the numerical amount of our population, their varied occupations and the fundamental sources of their industry and wealth, depend, in a great degree, upon the geological character of the strata on which they live. Their physical condition also, as indicated by the duration of life and health, depending on the more or less salubrious nature of their employments; and their moral condition, as far as it is connected with these employments, are directly affected by the geological causes in which their various occupations originate."

William Buckland, 1836

## GEOLOGY AND TIME TRAVEL

"The train soon quit the flat Thames Valley beyond Reading, and with it the soft sands and hard cobbles of the Reading Beds, laid down when there were mammals and birds on land, and crabs in the sea, and when the world would have felt familiar. Then I was speeding over the Chalk, and back in time to the company of dinosaurs. The train travelled on towards Bath, where Jurassic limestones and shales take turns across the countryside, the former proud with ancient corals, the latter dark and low, with ichthyosaurs and plesiosaurs, 'sea lizards' that grasped Jurassic fish and ammonites. Of ammonites, plesiosaurs and ichthyosaurs, nothing remains but their fossils. Under the River Severn into Wales, and I was back *before* the time of the dinosaurs, to a time when Wales steamed and sweated, not with the fires of smelting, but with the humid heat of moss-laden and boggy forests in coal-swamps, where dragonflies the size of hawks flitted in the mist; and then on back still further in time, so far back that life had not yet slithered or crawled upon the land from its aqueous nursery."

Richard Fortey, 1993

chapter twelve

# Fossils, Debts And Friends in High Places

## *The sale of William Smith's collection*

### PROLOGUE

The year is 1815. Great Britain has emerged victorious from a long and exhausting war against the French. The ingenuity of its inventors and engineers, and the abundance of its mineral resources, has given the nation a head start in the Industrial Revolution. Gentlemen of private means are beginning to take an interest in the science of geology. But those geologists, surveyors and mining engineers drawn from the trading classes must continue to sell their expertise in the market-place. They must pay for their own publications, so they do not publish anything. Information is not exchanged and the science of geology is in danger of fragmentation and stagnation. Meanwhile in Germany and France, the *Bergakademien* (Mining Academies), the *Ecole des Mines* and government agencies pay handsome salaries to mineralogical surveyors, and undertake to publish any geological maps that their surveyors produce. Thus, despite the individual brilliance of its men of science, does Britain begin to lose its advantage over these countries.

THE PARTS

*Principals*

WILLIAM SMITH, mineralogist, geologist, surveyor and engineer

WILLIAM LOWNDES, Chief Commissioner of the Tax Office, amateur botanist and geologist, Fellow of the Geological Society of London

THOMAS HOBLYN, Senior Clerk in the Treasury, also scientist and inventor

CHARLES HATCHETT, chemist and mineralogist

JOSEPH PLANTA, Principal Librarian and head of the British Museum staff

CHARLES KONIG, Curator, Natural History Collections, British Museum

*Other parts*

JOHN PHILLIPS, Wm Smith's young nephew, later to become Professor of Geology at Oxford and President of the Geological Society

NICHOLAS VANSITTART, Chancellor of the Exchequer

JAMES BROGDEN, MP

SIR JOHN THOMAS STANLEY

BARNE BARNE, MP

SIR JOSEPH BANKS, President of the Royal Society

GEORGE HARRISON, CHARLES ARBUTHNOT, officials at the Treasury

## I

*William Smith, having been a successful canal and mining engineer and surveyor for some twenty years, has fallen on hard times. Having acquired the lease to land at Combe Down, Somerset near to his own estate at Tucking Mill, Mr Smith raised money to build a small railway to link the quarries at Combe Down to wharves on the nearby Somerset Coal Canal. Unfortunately the quality of the stone (which he, above all men, should have been fit to judge) proved poor, and Mr Smith was left with considerable debts. During these twenty years Mr Smith has followed an interest in the rock strata through which his canals and mine workings have been dug. He has spent energy and yet more money on compiling a Mineralogical and Geological Map of the whole of England*

*and Wales, the first that was ever made, which is so well advanced that its publication is imminent. Unfortunately, again, the costs of printing and coloration have proved so great that the map is likely to cost Mr Smith more money than it can make him. In the course of his work Mr Smith has also collected a vast number of fossils from locations all over the country. These are classified into species and are able to be catalogued to their precise locations and strata. Mr Smith's plan, indeed his only chance of salvation from the debtors' prison, is to sell his fossil collection to the government. But first he must convince them of its worth.*

*Portrait of William Smith painted by H. Forau in 1837.*

1 July 1815
*Wm Smith's diary.*

Mr Vansittart, Chancellor of the Exchequer and Mr Brogden, Chairman of Ways and Means, called with Sir John Stanley, Bart., to view my Map of strata and collection, and proposed to purchase the latter for the British Museum. They advised me to draw up a memorial to the Treasury.

19 July 1815
*Extract from Memorial received by Treasury on 19 July 1815, from Mr Smith, recorded in Treasury Register of In-Letters as* Mr Smith. rel. to purchase of his

collection of Fossils for the British Museum, *in which Mr Smith stresses the value of his work to the nation and the financial burden of publishing his maps and papers, but does not mention his financially disastrous venture in Somerset.*

The pursuits in which I have been so long engaged have not been attended with advantage to myself of sufficient importance to enable me without assistance to carry into Effect the publication of the Papers I have written Illustrative of my Plan.

I have therefore taken the liberty of requesting Your Lordships consideration of my Case with reference to the purchase of my Collection for the British Museum which would add very much in point of utility to the Collection now deposited there and would moreover I trust afford me means of giving to the Public a Mass of information which would be found highly interesting and beneficial to almost every Class of Society.

15 Buckingham Street
York Buildings

29 September 1815

*Mr Smith has had no reply from the Treasury, and is now desperate to satisfy his creditors. He writes to Thomas Hoblyn at the Treasury.*

Sir

I am truly sorry to be under the necessity of troubling you with my complaints of a pecuniary nature but the great Expense which I have incurred in prosecuting my discoveries of the Strata and in bringing out the Map is likely to be attended with ruinous consequences to me and unless I can obtain some assistance my Interest in the Map must be immediately sold to satisfy the very urgent demands of my Creditors. If the Lords of the Treasury could be prevailed upon to make me an advance of five or seven hundred pounds on account of my Fossils proposed to be purchased for the British Museum, which should be deducted out of the Sum they may hereafter be pleased to give me for them, it would render me a most essential service and enable me to proceed with the Work. I am most respectfully

Sir

Your obliged Servant

15 Buckingham Street

York Buildings

*Wm Smith's diary, same day.*

Mr Hoblyn of the Treasury called to inform me that if 5 or 700£ advanced on my fossils proposed to be purchased for the British Museum would relieve my difficulties, he thought the Treasury might be disposed to advance such a sum.

3 October 1815

*A Meeting of the Treasury Commissioners considers Mr Smith's Memorial, perhaps under pressure from Mr Hoblyn.*

*Treasury Minute Book record of meeting of the Lords Commissioners.*

Oct. 3rd. Read memorial of Mr Wm Smith Engineer and Mineralogist transmitting a Geological Map exhibiting a general view of the numerous strata of which this Island is composed and offering for Sale a Collection of Fossils, etc, for the British Museum being collected specimens of each stratum. My Lords having taken into consideration the great importance of adding to the valuable Collection of Organic remains now deposited with the British Museum more particularly of those specimens connected with the Geology of this Country are pleased for the purpose of ascertaining the Value of those Specimens now offered for Sale by Mr Smith 15 Buckingham Street to request Chs Hatchett and J Planta Esqs to take the trouble of investigating the same and reporting to My Lords their Opinion of its merit and what Sum they conceive would be a fair consideration for the purchase thereof.

*Wm Smith's diary, same day.*

E[vening]. Thomas Hoblyn, Esq., called from the Treasury to inform me he had written, by the Chancellor of the Exchequer's desire, to Mr Charles Hatchett and another gent. (Mr Planta) of the Museum to come and value my collection and that they would do immediately.

4 October 1815
*Wm Smith's diary.*

Began cleaning the fossils ready for inspection of the aforesaid gentlemen. Mr Koenig, of the Mineral Department of the British Museum, called to have a Map of Strata fitted up for the Museum.

5 October 1815
*Wm Smith's diary.*

Mr Planta (the principal of the Museum) called to inform me Mr Hatchett and himself had agreed to come and value the Fossils at 12 o'clock on Monday next.

10 October 1815
*Wm Smith's diary.*

Messrs Planta and Hatchett came and viewed the Fossils and for want of a catalogue agreed to report generally to the Treasury on their value to the Museum. Mr Koenig with them.*

* Smith, having a prodigious memory, had no need of a personal catalogue, nor had he yet had time to prepare one for the museum. Hatchett was principally an analytical chemist; Planta was an administrator, not a natural historian; and Koenig, whose name is also spelt 'Konig' in these documents, undoubtedly knew much less about the importance of fossils than Smith.

13 October 1815

*Mr Smith becomes anxious about the progress of the report, revealing his ever pressing financial concerns.*

*Wm Smith's diary.*

Called on Mr Hoblyn at the Treasury Chambers to enquire if Messrs. Hatchett & Planta had made the report on my Fossils.

17 October 1815

*Report by Messrs Hatchett and Planta in the form of a letter to Mr George Harrison, Secretary to the Board of the Treasury, in which they express interest in Mr Smith's collection, but disapproval of its condition. Though not well acquainted with the science of fossils, they perceive that the value of the collection resides in its precise cataloguing, so that the locality and stratum of each specimen can be distinguished.*

Sir

In consequence of the Letters you honored us with on the 3rd Inst: respecting Mr Smith's Collection of Organic Remains, we, together with Mr Konig, the Keeper of the Department of Nat: Hist: in the Museum, whom we wished to join with us in this inquiry, took the earliest opportunity of viewing the said Collection.

Finding that it was not in the State in which it is meant to be in order to enable us to form any approximate valuation of it, we have desired Mr Smith to supply us with such data as we deem necessary for the purpose: and this he has promised to do without delay.

Meanwhile however we think ourselves justified in complying with his request to give you the previous information that by what we have seen of the Collection, though of itself to be of no great value, yet connected as it is with objects that may prove of great public utility, appears to us very proper to be deposited in the National Repository; and that, he having apprized us that he stands in need of some immediate assistance, we are of Opinion that an imprest of a sum not exceeding One hundred pounds may be granted to him without any danger of detriment on the part of Government.

We have the honor to be
Sir your most obedient humble Servants
J. Planta
Charles Hatchett

17 October 1815

*Treasury minute recording acceptance of Messrs Planta and Hatchett's report.*

Give directions to Mr Speer to pay to Mr Smith the sum of £100 as of special service and let the same be deducted from the amount which may be awarded to him as the value of his Collection in case it should be purchased on the part of the Public.

20 October 1815

*Wm Smith's diary.*

Waited on Mr Hoblyn at the Treasury and recd of him £100 in advance (especial services) toward the sum to be paid me for my Organic Remains which are to be fitted up in the same manner at the British Museum.

21 November 1815

*John Phillips, William Smith's 14-year-old nephew, arrives to help his uncle in arranging his collection.*

23 November 1815

*Wm Smith's diary.*

With JP [John Phillips] arranging shells according to Linnaeus.

3 December 1815

*Mr Smith's progress in cataloguing his collection is demonstrated by a complete 'Table of the Genera of Organic Remains in the Different Strata'. This is shown to an appreciative audience at Sir Joseph Banks's Sunday evening 'Conversazione'. Progress with the Museum and Treasury seems to have ground to a halt, with the museum officers showing little interest in following up their earlier visit, but Mr Smith is still in urgent need of funds. He lobbies his friends and contacts, hoping to gain influence at the Treasury.*

*In early December Mr Smith takes copies of his 'Table of the Genera of Organic Remains' to his friends Mr Barne Barne and Mr James Brogden.*

13 December 1815

*Mr Smith calls on Mr Hoblyn at the Treasury and presents him with a copy of his 'Table of the Genera of Organic Remains'.*

14 December 1815

*Wm Smith's diary.*

Went to Mr B. Barne and then to Mr Lowndes at the Tax Office and advised with him on my obtaining more money from the Treasury.

*Mr Lowndes, one of the most powerful men in the government, and a keen geologist, is also presented with a copy of the 'Table of the Genera of Organic Remains'.*

15 December 1815

*Wm Smith's diary.*

[Spent day] drawing up arguments for obtaining more money, for Mr Lowndes to send to the Chancellor of the Exchequer – and attended with same.

21 December 1815

*Mr Lowndes's efforts produce an immediate effect, and Mr Smith is informed by Mr Barne that Mr Hatchett will be calling on him again.*

22 December 1815
*Wm Smith's diary.*

At home all day expecting Gentlemen to call respecting Fossils for the British Museum.

23 December 1815
*Wm Smith's diary.*

Mr Hatchett Mr Planta and Dr Konig called and viewed Fossils their additions and improvements and desired me to report to Mr Harrison of the Treasury my progress and when they would be ready to remove to the British Museum.

*Further report to Mr George Harrison of the Treasury from Messrs Hatchett, Planta and Konig, in which they estimate the total value of Mr Smith's collection at a low price – less than he originally hoped to receive as an Advance.*

Sir

In consequence of some further intelligence received from Mr Smith respecting his Geological Collection, we this Day returned to his House in Buckingham Street, viewed the Additions that have been made since our former visit, and received from him such further information as he could give us on the subject.

The result of this investigation, though it be not yet so satisfactory as we could wish, yet leaves us no doubt as to the propriety of the Opinion we delivered in our former Report of the 13th Oct. last . . . And being moreover desirous to promote the intended purchase as much as may be consistent with our duty, and at the same time to relieve Mr Smith from the embarrassment under which we are told he presently labours; we have agreed to propose the following progressive payments.

1. That whenever Mr Smith shall report to you that he is ready to begin the removal of the Collection to the British Museum, The further sum of One hundred pounds be impresseded to him.

2. That as soon as the whole shall be so deposited, another hundred pounds be paid to him – &

3. that when he shall have completed the arrangements to the satisfaction of a certain number of Persons conversant in that science, he be entitled to the additional sum of Two Hundred pounds making in the whole (with the addition of the hundred pounds already imprested to him) a total Remuneration amounting to Five hundred pounds.

25 December 1815

*Mr Smith's sense of urgency was so great that even Christmas Day could not be spared.*

*Wm Smith's diary.*

All day arranged fossils. Cleaning and better numbering Specimens to go to the Museum.

28 December 1815

*Mr Smith writes to George Harrison at the Treasury, as requested by museum officers.*

Sir,

Mr Hatchett & Mr Planta having viewed the addition to my Collection of Fossil Organic Remains which I have collected and arranged for the purpose of explaining the nature and order of the British Strata, and desired me to report to you my progress therein.

I am happy to inform you of a great improvement in the Collection by which the locality of each specimen is readily distinguished.

A sufficient number of them, in each Stratum are already cleaned and re-marked for fitting up in the same order in the British Museum as soon as I may receive Instructions for their removal.

I am, etc.

29 December 1815

*Minutes of the meeting of the Treasury Board.*

My Lords concur in the recommendation of Mr Planta and Mr Hatchett and are pleased to direct a letter to be written to Mr Smith acquainting him with the time and manner in which their Lordships propose to pay him for this collection.

*Mr Smith's diary, in which he records that Mr Lowndes, a man of great influence, is unhappy with the museum officers' report.*

[Visited Mr Lowndes who] sent to explain to me Mr Hatchett's report to the Treasury, expressed his disapprobation thereon & said he would write to the Chancellor.

2 January 1816

*Wm Smith's diary.*

[Visited Mr Lowndes who] read me particulars of Mr Hatchett's report and said I had better wait the result of further reference.

3 January 1816

*Wm Smith's diary records that he had now received official notice from Mr Geo. Harrison at the Treasury accepting Mr Hatchett's proposals.*

Received official account from the Treasury of the Sums to be advanced for fitting up my Geological Collection in the British Museum.

4 January 1816

*Mr Smith, though undoubtedly disappointed with the sum proposed, is anxious to receive the second portion – in his situation swiftness of receipt must override the possibility of negotiating for a higher sum. But there are more difficulties in store for Mr Smith in his dealing with the British Museum. He may be ready to dispose of his Collection, but are they ready to receive it? He writes to the Lords Commissioners of the Treasury.*

My Lords
Being honored with Your Lordships' commands as to the disposal of my Geological Collection, I have the satisfaction of reporting, in conformity thereto, that I am this Day ready to begin the removal of the Fossils to the British Museum and that some boxes are already packed for that purpose.

*On the same day his diary reads:*
Wrote Report to Lords of the Treasury on the delivering of my Fossils, and attended with same at the Treasury. Saw Mr Hoblyn and carried Report to Register Room. He said 100£ was ordered me and I should be sent for.

8 January 1816
*Wm Smith's diary, in which his growing exasperation is evident.*
[Called on Mr Lowndes] to know if anything had transpired – nothing.

9 January 1816
*Mr Smith calls twice at the Treasury without result. The Board did however meet on this day, as the Treasury minute records.*
Jan 9th. Read letter from Mr Smith respecting his geological Collection and stating he is ready to begin the removal of the Fossils to the British Museum. Mr Lushington states to the Board that directions have been given to Mr Speer to pay to Mr Smith the sum of £100 in further payment of this collection.

10 January 1816
*Wm Smith's diary records the generosity of Mr Lowndes.*
Rec'd of Mr Lowndes who is authorized to re-take it of the Treasury sum of 100£.

*Mr Smith's accounts record that £42 was required to pay his half-yearly rent and rates, while £25 went to settle an account in Bath, leaving only a few pounds towards his living, and the repayment of further debts. His desperation, though eased a little, is ever-present.*

12 January 1816

*Now Mr Smith's attention turns to the next portion of money, due on receipt of the collection by the British Museum. Once again Mr Lowndes is a powerful ally in demonstrating that any delay is caused, not by Mr Smith, but by the museum authorities. Mr Smith's diary reads:*

Went to Mr Lowndes and BM to enquire if room appropriated to my Geological Collection was prepared.

*Evidently it was not.*

20 January 1816

*Mr Smith attempts to reduce the delays by acting as messenger and advocate between the Treasury and Museum, aided as ever by Mr Lowndes. His diary breathlessly records:*

Morning. To Mr Lowndes Somerset House from him to Mr Speer & Mr Harrison at the Treasury & back with message to Mr Lowndes and to British Museum obtained from Mr Konig answer to my note respecting time they would be ready to receive my Geological Collection.

*The note that Mr Smith badgered out of Mr Konig, Keeper of Natural History at the British Museum, shows the beginnings of that Institution's irritation at Mr Smith's demands, and at his use of influential friends to execute them.*

Dear Sir

I have to acquaint you, in answer to your note of this day, that the Apartment destined for the reception of your Geological Collection, & which is now under repair, will be ready in less than three weeks.

I am, Sir,
Your obedt. Sert.
Charles Konig
British Museum

22 January 1816

*The following Monday Mr Smith, as ever in desperate need of immediate funds, sends Mr Konig's note to the Treasury with a covering letter.*

Having in conformity to Your Lordships request reported my readiness to deliver my Geological Collection, and receiving no Instructions thereon, I applied to the officers of the British Museum who returned me the answer inclosed; and in consequence of the delay which the reparation of such apartments may require I am apprehensive that much valuable Time may be lost.

I must therefore humbly beg Your Lordships' consideration of the many difficulties I have already encountered in bringing such valuable information before the Public and that the further sum of one hundred pounds may be impressed to me to meet in some degree my present urgent necessities.

Wm. Smith
15 Buckingham Street
York Buildings

23 January 1816

*The Treasury Commissioners show themselves more alert than the British Museum to the plight of the renowned geologist. The Treasury minute for this day records:*

Read letter from Mr Smith dated 22nd inst . . . Under these circumstances My Lords are pleased to direct Mr Speer to advance to Mr Smith a further sum of £100.

*Mr Smith's diary records a visit from Mr Lowndes and Mr Barne, anxious that Mr*

*Smith should be able to demonstrate that his collection was entirely prepared for public use.*

[They] desired me to proceed with a Book of reference to the Specimens of Geol. Collection that they may be ready for the Museum when the rooms are repaired.

24 January 1816

*Mr Smith receives £100 from the Treasury. He immediately sends £20 to his brother in Bath, and pays over a further £55 in settlement of a legal action against him, which we may assume was for the settlement of an outstanding debt. Even these sums, though, were enough only to keep his creditors temporarily from his door. Mr Smith has now received £300, with the remaining £200 to be paid when he has 'completed the arrangements to the satisfaction of a certain number of persons conversant in that science'. By this we understand the arrangement of his collection within the museum – but first it must get there.*

*Mr Smith spends the next few weeks cataloguing his fossil collection, while also checking copies of his extraordinary Geological Map, as they arrive from the colourists.*

26 January 1816
*Wm Smith's diary.*

Making out catalogue and description of species of Echini in the Chalk Stratum.

5 February 1816
*Wm Smith's diary.*

Making out Cornbrash & Forest Marble Fossils with JP.

8 February 1816

*Wm Smith's diary.*

About catalogue of Great Oolyte & Fuller's Earth Fossils – for the Museum.

10 February 1816

*Wm Smith's diary.*

All day arranged Fossils & sorting and writing and making out particulars of localities Genera & Species in the under Oolyte.

7 March 1816

*Six weeks have now passed since Mr Konig's note stating that the room allocated for Mr Smith's collection would be ready in 'less than three weeks'. Mr Smith presses the Treasury once again, with a letter to the Lord Commissioners.*

My Lords

Since the last allowance made to me I have been occupied many weeks, to the entire interruption of my professional pursuits, in making out a Catalogue of my Fossils. They comprise 693 species and nearly 3000 Specimens of Organic remains; besides specimens of the strata, in which they are imbedded. The whole are now packed in cases ready for removal, but as the room at the Museum is not prepared for them, I hope your Lordships under these circumstances will be pleased to allow me to receive one hundred Pounds more of the five hundred Pounds awarded for the Collection.

15 Buckingham Street
York Buildings

12 March 1816

*Mr Smith follows up his letter of the previous week, and also prepares a document detailing the work which he has carried out on his collection, mentioning both its extensive nature and its value to the public.*

1 April 1816

*Mr Wm Smith has received no reply from the Treasury, but instead receives the following note from Dr Charles Konig at the British Museum.*

Dear Sir

Mr Planta desires me to acquaint you that he wishes to confer with you and me, on the best mode of placing and exhibiting your Collection in the Museum.

If you can make it convenient to call to-morrow, Tuesday, between the hours of 2 & 3 you will be sure to find him & I shall make it a point to be there at the same time.

I am
Dear Sir
your very obedt. Sert.

*We must assume that this meeting took place, and that the relevant plans were made. Mr Smith went off to Swindon in late April to work on the water supply for the Wiltshire and Berkshire Canal. He had no further correspondence from the British Museum until June. In the meantime he continued to publish works that were the foundations of the science of geology.*

1 June 1816

*The first part of William Smith's* Strata Identified by Organized Fossils *is published, as John Phillips later described, 'in consequence of an arrangement by which William Lowndes Esq. of the Tax Office, a very strenuous and judicious friend of Mr Smith, advanced £50 to pay for the first number'.*

8 June 1816

*Wm Smith's diary.*

Mr Koenig of the Museum came to say they wished to have the Geological Collection delivered.

10 June 1816

*Wm Smith's diary.*

Began delivering my Geological Collection to the British Museum.

18 June 1816

*Wm Smith's diary.*

Took to the Museum the remainder of Collection.

19 June 1816

*Wm Smith's diary.*

Took to the Museum two boxes omitted.

20 June 1816

*Wm Smith's diary at last records the settlement of the final portion of his money due.*

200£ rec'd of the Treasury (the remainder of five hundred awarded for my Geological Collection deposited in the British Museum).

*Thus the story ends – almost.*

## II

*As Mr Smith's debts continue to mount, he decides to press for more funds from the Treasury, through the Museum, on account of the extra work involved in preparing his collection for the museum. His requests are not well received – in some quarters. Here we see the value, or more truly the indispensability, of Friends in High Places.*

12 July 1816

*Letter from Wm Smith to the Trustees of the British Museum.*

Being on 2nd of Jany. last officially commanded by the Lords of the Treasury to remove my Geological Collection to the British Museum an application was made to the proper officer to ascertain the apartment destined for their reception, but the room being then under repair I was advised to proceed with the arrangement in my own House.

*Mr Smith describes the ways in which the arrangement of his collection is superior to any previously attempted, in Britain or on the continent, and continues:*
I have reason to hope that the Honble Trustees will not let these labors pass unrewarded, which on the Continent were but partially accomplished by the most able professors, appointed to investigate their Country at the expense of the nation.

*Mr Smith describes how his intricate, but necessary, numbering and labelling system . . .*
made it an arduous Task, for which I hope to be remunerated, over and above the value of the Collection, and for which this application is most respectfully made.

7 August 1816
*His Majesty's Treasury receives a similar, but yet more detailed letter from Mr Smith, entered in the In-Letter Register as:*
Mr Smith for remuneration for extra labour in completing his geological collection.

3 September 1816
*Mr Smith's letter and memorial to the Treasury are considered at a meeting of the Board.*
Read letter from Mr Smith praying remuneration for extra labour in completing his Geological Collection in the British Museum. Transmit Mr Smith's memorial to the Trustees of the British Museum . . . request they will favour My Lords with their opinion on the merits of his application.

23 September 1816

*Private letter from Mr Planta (British Museum) to Mr Harrison (Treasury), in which the administrator's feelings are espoused.*

Sir,

The Trustees of the British Museum having adjourned their Meetings to the 9th of Novr. next, it falls to my share to acknowledge the Delivery into my hands of your Letter to Them of the 13th inst: . . . Not being authorized to make any official answer to the demand of their Lordships respecting the contents of that Memorial, I can only mention as an individual that I am not a little annoyed at Mr Smith's fresh application, as in my opinion there is no ground for it, he not yet having completed the arrangement of the Collection, which was one of the express Conditions of his Agreement, & without which the purchase can be of no manner of use—

The plea on which he founds his further demand, viz. that he has made considerable additions to the Collection, it is not in my power to countenance, there being nothing to convince me of the truth of the allegation. As far as the inspection of Mr Konig and myself goes, we are not aware of any.

I keep the Memorial by me, in case you should desire me to lay it before the next meeting of our Trustees: but should you think this, on account of the distance of that meeting, unnecessary, I will return it without delay

I have the honor to be
Sir Your most obedient humble servant
J. Planta

15 October 1816

*Wm Smith's diary.*

Went to Mr Lowndes, prevailed on him to go to the British Museum to view and report on my Geological Collection, met him there with

Mr Planta and Mr Konig, agreed that fixture cases are wanting. Returned and drew up my Report, called and showed it to Mr Lowndes who added to it. Attended Mr Planta at Museum & got him to go to Mr Lowndes.

16 October 1816

*Letter from Mr Lowndes to Mr Planta.*

Dear Sir

I have read the enclosed Papers, containing Mr Smith's Representation of the extraordinary Trouble he has undertaken, since the valuation of his Collection, in arranging the several specimens of which it consists so as to render it usefull to the public, and I can with truth bear Testimony to the motive which actuated him and to the incessant Labour for many months in forming the Catalogue raisonne which to him was an entirely new Study . . .

I can also with equal Truth assure you that from the first Mr Smith considered his Labours as not within the Terms of his Sale, but undertook it in full Reliance that he would be remunerated for his Time & Trouble & knowing the Justice with which Government acts to those employed under it, I could but assure him would be the case.

I have the Honor to be
Dear Sir
Yr very faithful Sevt
W. Lowndes
Somerset House

*Among documents enclosed is one signed by Mr Smith wherein he advocates the economic value of the geological science which he is advancing – no doubt to contrast the sums mentioned with his own, as he sees it, meagre recompense from the museum. This is an extract.*

These specimens with many others which I have collected are of the utmost Geological importance as they completely prove the false principles on which a worthy Baronet was lately induced to sink for Coal in these Strata [at] the ruinous expense of 20,000£. The specimens which I can produce will for ever settle the false notions of Coal in the Strata beneath the Chalk which have long prevailed and prevent such wanton expenditures of money in search of it as have been the ruin of many.

Wm Smith
Civil Engineer & Mineral Surveyor

*Mr Lowndes not only seeks remuneration for Mr Smith's additional labours, he recommends that Smith should receive payment for any future work which would be of value to the nation.*

17 October 1816

*A letter from Mr Planta of the British Museum to Mr Harrison at the Treasury, enclosing Mr Lowndes's letter and Mr Smith's Memorial of 16 Oct. Mr Planta has, so to speak, changed his tune. Has his sudden enthusiasm for the value of Mr Smith's collection and work been aided by the intervention of Mr Lowndes, or has he come to the science of geology through some inspirational means?*

Sir

At the earnest request of Mr Wm Smith, I take the liberty of transmitting to you the inclosed statement of his past labours in the prosecution of Geological enquiries, & of the further services he has it in contemplation, if kindly encouraged, to render towards the furtherance of that Science. I should not however have thought myself justified in making this voluntary representation, did I not feel myself in some measure authorized by the inclosed Letter from Mr Lowndes, which strongly indicates how much he is impressed with the importance of the case, and how much he has the improvement of that branch of knowledge at heart.

I have the honor to be
Sir
Your most obedient hble Servant
J. Planta

*Note the use of the word 'voluntary', and remember that Mr Planta is a non-scientific officiary – he can have no view on the scientific value of Mr Smith's work.*

23 October 1816

*Mr Smith grows anxious, and remembers that, up until now, no Sum has been specified for his 'extra' work in arranging his collection. He writes to the Treasury. After a short description of the preceding events, Mr Smith comes to the point.*

I hope I may be allowed 100 or 120£ on account of what I have done, and may be further enabled to do, in which utmost endeavours shall be exerted to merit your kind patronage.

I have the honor to be most respectfully
Your very humble and obliged Servant
Wm Smith

25 October 1816

*Treasury minute.*

Read letter from Mr Smith of 23rd inst applying for some remuneration for his trouble in arranging the minerals lately purchased for the British Museum.

Let Mr Speer pay £100 to Mr Smith in consideration of his trouble in this Service.

26 October 1816

*Letter from Mr Charles Arbuthnot, Secretary to the Treasury to Mr W. Lowndes, which reveals that the previous day's decision was abetted by further work by that honourable gentleman and friend to Mr Smith.*

My Dear Sir,
I have received your Letter in favor of Mr William Smith, & as I am on the point of quitting Town I have only time to say that Mr Vansittart was with me when your letter came & he took it with him to the Board promising to pay every attention in his power to it. Excuse the haste in which I write & Believe me

I am
Yours most sincerely
C. Arbuthnot

## III

*In which Mr Smith must sell more of his Collection in order to avoid the Debtors' Prison.*

January 1818

*Letter from Mr James Brogden, MP to Mr Vansittart of the Treasury Board. This follows letters from Mr Smith to the Treasury offering more of his collection of fossils for sale, and an approving report on them by Mr Konig of the British Museum, aided by the very expert opinions of Sir Joseph Banks, President of the Royal Society.*
*We may gauge Mr Smith's predicament from the tone of the letter.*

Dear Sir,
Poor Smith's distress and importunity force me again to intrude this unpleasant subject on your notice. The annexed paper contains a statement of what he thinks are his claims for some compensation for the time and labour employed in the construction of his map. He informs me that Elkington for his system of drainage obtained the Parliamentary grant of £1000. He has now before the Treasury (1 month) a claim for the value of another collection of fossils deposited in the British Museum. If you can settle his claim to-day

and make a small advance on it you will save him from arrest and enable him to resume his useful employment in the country.

I am
dear Sir
Your faithful Servt.
Jas. Brogden
Parl. Liby.

*A brusque letter from Charles Arbuthnot to George Harrison at the Treasury.*

Dear Harrison
Smith the Map Maker (I mean the Strata & Fossil Man) hopes to receive from the bounty of the Treasury some aid; I heard last night in the H. of C. that if he don't get that aid he will be arrested tomorrow [underlined 3 times].

*As a result of these letters Mr Smith receives an advance of £30 from the Treasury. The following month he received notice that he was to be paid the balance.*

12 February 1818

*While absent in Monmouthshire on business, Mr Smith receives the following letter from the Treasury.*

Sir
The Lords Commissioners of His Majesty's Treasury having had under consideration your Memorial of 20 August last relative to certain additional Fossils collected by you in the course of the last year – I have it in command to acquaint you that They have been pleased to direct Mr Speer of this office to pay you One Hundred Pounds for the Purchase thereof on the production of a Certificate from the proper officers of the British Museum that the said Fossils have been duly delivered into the possession of that Institution.

I am Sir

Your most obedient Servant
Geo. Harrison

28 March 1818

*Mr Smith collects the balance of £70 from the Treasury – the last payment he was ever to receive from that body.*

## EPILOGUE

The proceeds from these sales did not put an end to Mr Smith's financial difficulties. In 1819 he was forced to sell the lease on his London house, together with all the furnishings and most of his possessions. In 1820 he spent ten weeks in a debtors' prison. For the next four years he lived the life of a nomad, wandering through the North of England, accompanied only by his faithful and loving nephew John Phillips, and sustained physically and mentally by the conduct of that practical science to which he contributed so much. Finally he found rest and secure employment at the famous resort of Scarborough.

The Geological Survey was eventually established by the British government in 1835, just four years before William Smith's death, and geology began to be a profession.

Mr William Smith's Collection of Fossils was, after so much work and anxiety, never put on display in the British Museum building in Bloomsbury. When the natural history collections were moved to the new museum in South Kensington from 1880, Smith's collection went with them. In 1885 when the palaeontological galleries were opened, Smith's fossils were at last displayed in the manner which he had arranged – nearly seventy years after their purchase.

chapter thirteen

# The Fifth Reptile

## *Gentlemen and men of honour*

The fossilised skeleton of a massive *Plesiosaurus dolichodeirus,* found at Saltwick in the spring of 1841, became the subject of a bitter row between the Museums at Whitby and at Cambridge. Most of the correspondence has been preserved at Cambridge University Library, in the Adam Sedgwick manuscript collection. Relations between geologists in the two towns began in a friendly manner, but soon deteriorated. The plesiosaur first arises in a letter from George Young, cleric, amateur geologist and moving force behind the Whitby Museum, to Adam Sedgwick, Professor of Geology at Cambridge University.

Whitby, 2.9.1841

"My Dear Sir,

I send you inclosed a sketch of our Ichthyosaurus now on sale at Whitby. It is a good specimen & of considerable size, above 17 feet long, I think it was the property of the late Mr Hunton of this neighbourhood, & his friends value it at £70. This is certainly much above its value; but if you consider it suitable for your Museum, you could say how much you would give for it.

You have no doubt heard of the Plesiosaurus found here in the spring: we want to secure it for our Museum but will have to give at least £150 for it.

Should you have no wish to purchase the Ichthyosaurus, perhaps you will take the trouble to return the drawing; unless you wish to show it to friends in London.

With best wishes, I am,
My Dear Sir,
Yours truly
George Young"

By now the plesiosaur fossil was on public display above Matthew Green's workshop on Haggergate in Whitby. The owners were presumably earning some money from their find while they waited for an offer of purchase which met their expectations. In 1824, £7 had been enough to buy a perfect specimen of a fossil crocodile; by 1841 the market had changed.

J. W. Clark, Professor of Anatomy at Cambridge, and a close friend of Sedgwick, with a good knowledge of palaeontology, was travelling on the Yorkshire coast in the autumn of 1841. The new Woodwardian Museum had opened in Cambridge the previous year, and was in need of prestigious specimens. Clark wrote to Sedgwick from Whitby in a state of high excitement.

Whitby, 25.9.1841

"My dear Sedgwick

I do not know whether you are aware that about 4 months ago a long-necked Plesiosaurus was taken out of the Alum shale near Robin-Hood's Bay. I have seen the specimen. It now lies in the garrett of a cottage opposite our lodging here in Whitby: & is the property of three men who work the cliff, where it was found, for jet. It far surpasses anything of the kind I have ever seen. It is very nearly perfect – It is defective only in the scapulae where they ought to cover the ribs behind

# NOW EXHIBITING DAILY,

*In a Room over the Shop occupied by Mr. Matthew Green, Haggersgate,*

**A SPLENDID AND VERY VALUABLE FOSSIL**

# "PLESIOSAURUS DOLOCHODEIRUS,"

***Recently Found in Whitby Cliffs.***

This unparalleled Organic Specimen of so extraordinary an Animal measures 15 feet in length, and 8 feet 5 inches across the fore Paddles. The Neck is 6 feet 6 inches long, exclusive of the Head.

Among the multiplicity of Fossil Petrifactions discovered in the neighbourhood of Whitby, this by far surpasses all, even the famed Crocodile in the Whitby Museum; indeed it is questioned whether any fossil remains were ever discovered equal to that of this wonderful species of the Plesiosaurus tribe. The specimen is entire, without, we believe, a single joint wanting, and very cleverly excavated from the strata in which it was found. Among the notes appended to Goldsmith's Animated Nature, by Alexander Whitelaw, we find the following remarks in reference to this singular species: "Perhaps there has been no animal created of a more extraordinary form than the *Plesiosaurus Dolochodeirus*. In the length of the neck it far exceeds even the longest necked birds. It is in this species five times the length of its head; the trunk of the body four times the length of the head; and the tail three times; while the head itself is only a thirtieth part of the whole body. From the whole physiology of the Animal, Mr. Connybeare says, that it was aquatic is evident from the form of its paddles; that it was marine is equally so, from the remains with which it is universally associated; that it may have occasionally visited the shore, the resemblance of its extremities to those of the turtle, may lead us to conjecture; its motion, however, must have been very awkward on land; its long neck must have impeded its progress through the water, presenting a striking contrast to the organization which so admirably fits the Ichthyosaurus to cut through the waves. May it not therefore, be concluded, that it swam upon or near the surface, arching back its long neck like the swan, and occasionally darting it down at the fish, which happened to float within its reach."

**GENTLEMEN 6d. WORKING MEN 3d. CHILDREN 1d.**

*Whitby, August 7th*, 1841.—HORNE AND RICHARDSON, PRINTERS, WHITBY.

*Poster advertising the exhibition of the plesiosaur above Matthew Green's shop in Whitby.*

– & in the pelvis, also behind. The rest is beautifully complete and distinct. The specimen is 18 feet long. I counted about 42 cervical vertebrae – somewhere about 20 dorsal – and the caudal (I know not how many) also quite distinct. The four paddles are as perfect as if the live animal had been dissected – only that the humeri and femurs are a little broken at the upper ends. I cannot think that the existence of this specimen is very generally known – for when at Scarborough when I visited the Museum lately with Dr Murray & asked him what was to be seen at Whitby, he mentioned the fossil Crocodile of the Whitby Museum with high praise but did not say a word of this Plesiosaurus. The possessors of this noble specimen appear to know its value. They at first asked 500£ for it – but seem not to have had a bona fide offer above £150. This latter sum appears to have been offered by Dr Young of Whitby for the Whitby Museum – & has been declined. But such a treasure can not long be hidden: & if we would have it at Cambridge – where I hope to see it – something must be done quickly.

I have had the man with me this evening who owns the cottage where the specimen is. He is 1/3rd owner. I asked him – after due preliminaries – whether he would take 200£ for it. He at first said he would not: but after some parley he said he would consult his partners & give me an answer in a day or two.

I think it very doubtful that the thing will be bought for this sum. If it can I shall perhaps take the risk of the purchase upon myself. But the object of my letter is to request you to tell me immediately – for I leave Whitby on *Friday next* – whether you have heard of this specimen: & if you have what its character is with geologists, & what its value. And further I wish to persuade you, (because I do not think the owners will take the sum I have mentioned) to come down here without delay, if you have not heard of the specimen – & see it with your own eyes. You will caper round it with delight: – & then carry it away with you.

I trust this will find you in Cambridge for the Fellowship Examination. Do not miss a post in sending me a line & direct to me.

Post Office
Whitby
Yours my dear Sedgwick
always
J. W. Clark

Clark returned to Cambridge without hearing from Sedgwick. He evidently wrote to Whitby the following month, making an offer for the fossil. The fossil's owners wrote back to him immediately, and showed their preference for gentlemen of another town, above those they knew in their own.

"Dr Clark, Cambridge

Whitby Octr 27th 1841

Dear Sir
most agreeable to your letter which we recd Dated Octr 25 with the Sum Expressed 220£.
Dr Sir – we have now a better bid for the Plesiosaurus than you have carried forward with and if you Say 250£

then it will be at your demand – we shall waite your immediate answer on the same before we dispose of it. So if take it you will please to say which way you will have it forwarded. You will please to say if you pay the Expence of it coming to you –

Now Sir to give you a clear understanding into the affair – the Whitby Gentlemen are wishful to have [it] themselves – and we don't feel willing they should – So you will oblige us with your answer – we remain yours respectively

Matthew Green"

Clark must have replied to this letter immediately, and received another in reply, this time signed by all three owners.

"Dr Clark, Cambridge, speed to hand

Whitby, Nov 3rd 1841

Dear Sir

In reply to yours Dated the 29th Octr and have made up my mind for you to have the plesiorus for the Sum of £230 as Expressed in your letter – also by you defraying the Expenses of packing and boxes and the carriage of the Same as I wish to come along with it and see it Safe deliver'd without any damage – therefore Dear Sir – you will please to Send a little money when you write again and it will [be] ready to Set off by the time we receive your answer – to say which way you wish it to come Safe to you. So shall waite your immediate answer

and to remain yours

respectively

Matthew Green
Wm Brookbank
Abram Brookbank"

Before the arrangements were made to transport the specimen Dr Clark received the following letter from George Young and Richard Ripley of the Whitby Literary and Philosophical Society, owners of the Whitby Museum.

"To Dr Clark, Cambridge

Whitby, 6.11.1841

Sir

We are informed that you have bought the Plesiosaurus so long on sale here, and which we have been endeavouring to purchase for the Whitby Museum as appropriate companion to our great Crocodile. You are aware of our negotiating for the purchase, but surely you did not know how far the negotiation had advanced, otherwise you would scarcely have felt yourself justified in stepping between us and the prize which we had good reason to consider as our own.

We had resolved to give 200 pounds or guineas for it; two out of the three owners had agreed to our terms, and while we were waiting for the concurrence of the third, viz Matthew Green, whom accidental circumstances prevented us from meeting, it appears that he had been in correspondence with you, and on your offering him a little more, he had agreed to sell it. – Had a same fossil been discovered at Cambridge, & had we stepped forward and outbid you, while you were negotiating the purchase, you would have felt as we do now. – It is natural for you to be zealous for the interests of your Institution, as we are for ours, but we must not in our zeal, forget the position which we hold in society as gentlemen and men of honour.

Under all the circumstances you will, we trust, not judge us unreasonable, in hoping that you would decline the bargain in our favours: in which case we shall feel a pleasure in presenting your Museum with a correct cast of this valuable fossil. – Professor Sedgwick, who has always appeared to us to have a high sense of honour, will, we hope, be ready to acquiesce in our proposal. – Should we, however, be mistaken in our hope, regarding the decision that may be come at by you & your friends in this case, it would be some compensation for our painful disappointment, and abate somewhat of that sense of wrong, which in such circumstances it is natural for us to feel, if you will consent to let the Plesiosaur remain here till we can take a cast of it for our Museum. Care would of course be taken that no injury should be done to the specimen, and we should also remunerate the men for any trouble and delay it might cause them.

Waiting your friendly and equitable reply, we are
Sir,
Yours respectfully
George Young
Richard Ripley
Secretaries to the Whitby Lit. & Phil. Society"

Clark replied immediately, and in furious tones.

Cambridge, 8.11.1841

"Gentlemen

I this morning recd. a letter from you on the subject of my late purchase of the Whitby Plesiosaurus, which, on account of the grossness of the imputation which it casts upon me and the coarseness of the language in which it is couched, I had resolved at first not to notice. On further reflection, however, it occurs to me that persons, who could write that letter, will probably misinterpret my silence: and, instead of attributing it to indignant contempt of unmerited insult, may construe it as an admission that their charges have some foundation. I therefore enter into some detail as due to myself, & due also to those gentlemen connected with your society who appear to be indignant at this who I have no right to implicate with your conduct.* When I was at Whitby in the summer the Plesiosaurus was openly in the market for any purchaser who might choose to give the price demanded for it. The Directors of the Scarborough Museum had been in treaty for it as well as those of the Whitby Museum: & the proprietors of the fossil had declined the offers made by the parties connected with both institutions. The purchase had also been offered to the British Museum: and the agent who entered upon that negotiation was Mr. Ripley, one of the persons signing the letter to me. Further, when I was on the point of leaving Whitby, an offer larger in amount than any which had been previously made was

* This sentence is difficult to decipher on the original letter, being littered with deletions and amendments, a sign of Clark's evident fury.

advanced by myself, but was declined by the proprietors because they had then offered the fossil to a gentleman connected with the Dublin Institutions (who had been staying at Mulgrave) for £300, & he would have bought it without the slightest scruple (as he did other remains which he found in Whitby) had he thought it worth the money.

Putting therefore myself out of the question – it appears, that for the purchase of this Plesiosaurus, Scarborough, Whitby, London, Dublin & Cambridge have successively been in treaty. Yet in the face of these facts, and with the full knowledge of them (as I believe) you dare to tell me, in reference to my purchase, that I have stepped between you and the prize you hoped to gain, & to remind me "that we must not in our zeal forget the position we hold in society as gentlemen and men of honor".

In conclusion, I beg to inform you that, after your letter, you are in no position to ask a favor of me: & that I decline allowing any cast of the Plesiosaurus to be taken.

I am gentlemen
Your obt st
J. M. W. Clark"

While sending this letter to Whitby, Clark at the same time sent a short note to Sedgwick.

"Sir
I have had a most insolent letter from Messrs Ripley and Young accusing me of purchasing the Plesiosaurus

over their heads: which you know to be absolutely false. After making this charge, they ask me to allow them to take a plaster cast of the fossil. I write this to inform you that I will not allow any cast to be taken – & that if the application be made you will refuse to accede to it. Awaiting your answer to my last letter.

I am yr obt se"

Dr Clark received confirmation of the transport arrangements the next week in a letter from the Whitby shipping agent. The fossil was to go by ship from Hull to Lynn (King's Lynn), and then by barge up the Great Ouse to Cambridge. Presumably it had also come by ship from Whitby to Hull. It seems that Mr Green was not travelling with the fossil solely to see its safe arrival – he was going to Cambridge to get his money.

"Dr Clark, Cambridge

Whitby, 13th Novr 1841

Sir

Yours came duly to Hand Last night enclosing Post office Order for 10£ for which MG Partners are obliged.

M. Green started for Hull on Thursday and would leave Hull on Saturday for Lynn. So if all is well he would be in Lynn on Sunday. He had started two days sooner than your Letter as I thought it might be of consequence waiting another week as the Wether is favoured now and the Strenshalh would not have sail'd any more of [?] seven day Providing the Wether had been favourable, so I advanced him the money your friend Mr Stockdale would be holding out for him when the Lynn Steam boat arrived. I have no doubt and hope you will have him on Monday or Tuesday.

His partners desire me to say as Mr Green is no accountant will thank you to go to the Bank with him at Cambridge and get his Money Put in so as he can draw it when he gets Back to Whitby.

I here Enclose you the Pollacy of Insurance Effected at Newcastle which is more than I calculated on, but I have no doubt it is on the same terms as all respectable Insurance Compys are doing at this Season of the year.

PS you Perhaps in Receipt of this, will order Some person to Look out for the Barges as we calculate he [Green] would leave Lynn on Sunday or Monday Providing there was any Barges on them days sailing

Your most obedient Servant
for MG & Partners
John Authard"

To raise funds for the purchase of the Plesiosaur Professor Sedgwick decided to offer subscriptions to members of the University. Within days of the fossil's arrival in Cambridge he wrote to the Members of the Senate and other resident Members of the University.

TRIN COLL CAMBRIDGE
Nov. 19, 1841

"Gentlemen
As Curator of the Woodwardian Museum, I take the liberty of laying before you the following statement. In course of last summer a magnificent *Plesiosaurus* was dug out of the cliffs near Whitby, and was offered to the British Museum for £500. The offer was refused, not merely on account of the large price demanded, but

because the Museum already possessed an excellent series of specimens of that fossil genus. During the autumn Dr Clark (our Professor of Anatomy) saw the specimen while it was publicly exhibited and advertised for sale. Several large offers had then been made for it

*Adam Sedgwick, Professor of Geology at Cambridge from 1818 to 1873, in a mezzotint from 1833.*

from more than one quarter: as the only means of securing it for Cambridge, he made a still larger offer, and soon afterwards completed the negotiation, at his private risk, by purchasing the specimen for £230. I claim no divided honour with him in this spirited and liberal act; for the negotiation was begun without my knowledge, and I did not receive his letter, informing me that he had first offered £200 for the specimen, and meant if possible to purchase it for the University, till after my return to Cambridge. But I feel deeply grateful to him for what he has done; and I should have rejoiced to divide with him the responsibility of the purchase. There is certainly no place in the University so proper for the reception of the fossil, as the GEOLOGICAL MUSEUM. But in the present state of the Woodwardian surplus, nearly exhausted by the expensive fittings of the New Museum, I do not think myself justified in applying to the Heads of Houses, and requesting them to become the purchasers of the fossil out of the funds of which they are the Auditors. I therefore take the only step that is left; and venture respectfuly to call upon the Members of the University, who are interested in the honour of our Collections, to assist me by a subscription, and to enable me to give the fossil a permanent place in our Geological Museum. For the present it is deposited in the Lecture Room under the new wing of the Public Library; and may, on application, be seen by any member of the University.

I have the honour to be, Gentlemen,
Your very faithful Servant,
A. SEDGWICK"

Professor Sedgwick wrote to George Young about the affair at around this time. Using his personal prestige to repair the rift between the two towns, he could afford to be magnanimous – he, after all, had got the plesiosaur. Young was outgunned and forced to retract his earlier accusations, though he did not go so far as to apologise for making them. He was still furious that Whitby Museum should have lost a specimen that was discovered on their doorstep.

"Rev Profr A. Sedgwick DD, Cambridge

Whitby Nov 24 1841

My Dear Sir

The receipt of your letter of the 20th which came to hand last night affords me no small pleasure, as it clearly shows that if any wrong has been done to our Society as regards the purchase of the Plesiosaur, you have had no [hand] in it.

Any charges which we may have advanced against you were purely hypothetical – and it was with great reluctance that we could be brought to think you capable of doing any thing dishonourable – and now that the supposition which presented itself to our minds turns out to be erroneous, we cannot but rejoice that nothing has been done in this transaction that can in the slightest degree lessen that high esteem in which we have so long held you. – We were led to believe that the Fossil had been bought for the Fitzwilliam Museum for which you have made so many purchases, & knowing your connection with that Museum, we had no idea that Dr Clark, or any one connected with it, could make the purchase without your knowledge & concurrence. On that supposition your remarks made to Mr Ripley . . .

& to myself by letter, about the exhaustion of your funds & your not being able to venture on any large purchase, seemed utterly at variance with what had taken place. The facts which you state completely prove that we were wrong, so that not only your own conduct in the business, but Dr Clark's also, must now appear to us in a very different light, from that in which we first viewed it. – Indeed, our charges against Dr Clark were also hypothetical: we did not say (as far as I recollect) that he had acted dishonourably, but that if he had known our exact position in reference to the purchase, he would have seen the impropriety of stepping in between us & the prize, which we considered as ours. Two of the three proprietors had expressed their willingness to let us have the Fossil for 200 Pounds or Guineas, which we had resolved to give: through accidental circumstances we missed seeing the third, viz. Green, but learned from his wife that he would not part with it for less than 300 guineas. After a few days more had elapsed, when I called on Green, in the hope of completing the purchase, I then learned to my great surprise, that the Fossil was sold to Dr Clark. As we had not dropt our negotiation, during that interval, we could not consider the specimen "as fairly in the open market": at the same time we had much more reason to complain of the seller than of the purchaser.

At all events, Dr Clark should have made some allowance for the feelings of mortification and disappointment which at such a time we could not but experience. We may have expressed ourselves too strongly: & had he been in our circumstances, and we in his, it is likely he would have done the same. It is to little purpose to remark, that other institutions were endeavouring to secure the prize; you have too much candour to deny, that as it was a Whitby Fossil, we had a claim of preference to it above all other Societies. The friends of science have often urged it upon us, to devote our entire energies to the Fossil department of our collection, so as to have the best specimens preserved in the locality where they are found; & with that view, we anticipated the pleasure of placing this Plesiosaurus beside our great Crocodile. This object would have been attained had it not fallen into the hands of three low-minded fellows, so selfish and avaricious as to be incapable of feeling any thing like public spirit.

Allow me to add, that the scientific world owes a debt of gratitude to our Society; for had it not been for the impulse given to the search for Fossils in this quarter by its formation & its efforts, many excellent specimens, now enriching Museums & cabinets, would have remained undiscovered; & even this noble Plesiosaurus might have been lying quietly in its matrix. You are aware too, that we have been ready to serve other Institutions, by negotiating important purchases on their behalf.

Had the Plesiosaurus remained with us we intended making casts, with a view to supply various Societies and Public Institutions. Should you and your friends entertain the same thoughts, I trust that . . . Dr Clark will, on second thoughts, regard our Museum as not unworthy of obtaining a cast; that having lost the original, we may have a correct copy.

It is a consolation for us to know that this noble specimen is in the hands of those who will value & preserve it, & is placed where it will be accessible to thousands of scientific inquirers. –

With best wishes, I am, My Dear Sir,
Yours respectfully
George Young

P.S. We had a wish, in case we should find it impossible for us to raise the money for purchasing the Plesiosaurus, that it should be bought for the Bristol Museum. We had many inquiries regarding it to answer: & Mr Ripley, it appears, corresponded with some officers of the Museum on the subject."

Sedgwick's appeal for subscriptions was sent out to university members around the country who had an interest in natural philosophy. Being President of the Geological Society as well as a Cambridge professor, Sedgwick had extensive connections and influence (almost the whole Anglican clergy had connections to either Oxford or Cambridge). In ability to raise money, town museums like Whitby could not compete with an institution as powerful as Cambridge University. Among those who responded to Sedgwick's request was the Bishop of St David's in Pembrokeshire.

"Rev. Professor Sedgwick, DD
Cambridge

Abergwele
Decr 9 1841

Dear Sedgwick
Whatever the merits of your Monster may be, the modesty with which you plead his cause is irresistible,

and one cannot help feeling that a whole folio of good wishes would be but an inadequate acknowledgement for it. I feel it incumbent on me to do something more toward providing an honorable and comfortable establishment for the monster. But there is a difficulty which has caused me a good deal of perplexity and which after all I have not been able to solve in a satisfactory manner. A few weeks ago I was invited (as the French say) to subscribe to the restoration of the Round Church and I felt myself moved to contribute £5 to that object. Now just after this what is the proper amount of subscription for the relics of a Plesiosaurus?

I have tried the thing mathematically. Recollecting that I am a Fellow of the Geological Society [F.G.S.], I thought I might advance this proposition:

A Bishop : F.G.S. :: A Round Church : A Plesiosaurus

from which you get the equations

$$\text{A Bishop} \times \text{A Plesiosaurus} = \text{F.G.S.} \times \text{Round Church}$$

$$\text{Plesiosaurus} = \frac{\text{F.G.S.} \times \text{Round Church}}{\text{Bishop}}$$

But here I came to a stand briefly because I do not know what function of a Plesiosaurus a F.G.S. is. Perhaps life is too short to solve this problem, I have therefore determined to cut the knot, and by a motu propicio find the proportion in question at 5 : 3. Whenever you call for it £3 shall be paid to Mortlocks on account of the Magnifico.

Those who are able to judge of him with their own

eyes will it is to be hoped discard all such nice calculations and resolve to keep him – coute qu'il coute.

There is a tradition in this country that you were once thrown up by the sea on the coast of Pembrokeshire, and having been hospitably entertained, promised to come again, with your hammer, and break open some tracts which Murchison left untouched. I hope there is some truth in the legend, and that at some not very distant period you will redeem your pledge and at the same time refresh us with a sight of you here.

Meanwhile
Believe me
Dear Sedgwick [I am] Yours very truly
St Davids"

Not all the subscriptions came so readily. J. W. Clark wrote to Sedgwick in Norwich later that month. Professor Sedgwick was in the peculiar position of having to live part of the year in Norwich. His professorial stipend was too small to live on, and he was appointed a prebendary at Norwich in 1834. Occasional residence in the city was a condition of the lifelong post.

"My dear Sedgwick
I went to the Bank this morning and find your payment of £152-13s regularly transferred to my account. You are more delicate in this matter than accords with our agreement. For there is no laying of the money down on my part if you undertake to pay before you receive.

I learn from Anstead that the subscription proceeds. With the subscriptions attached to some

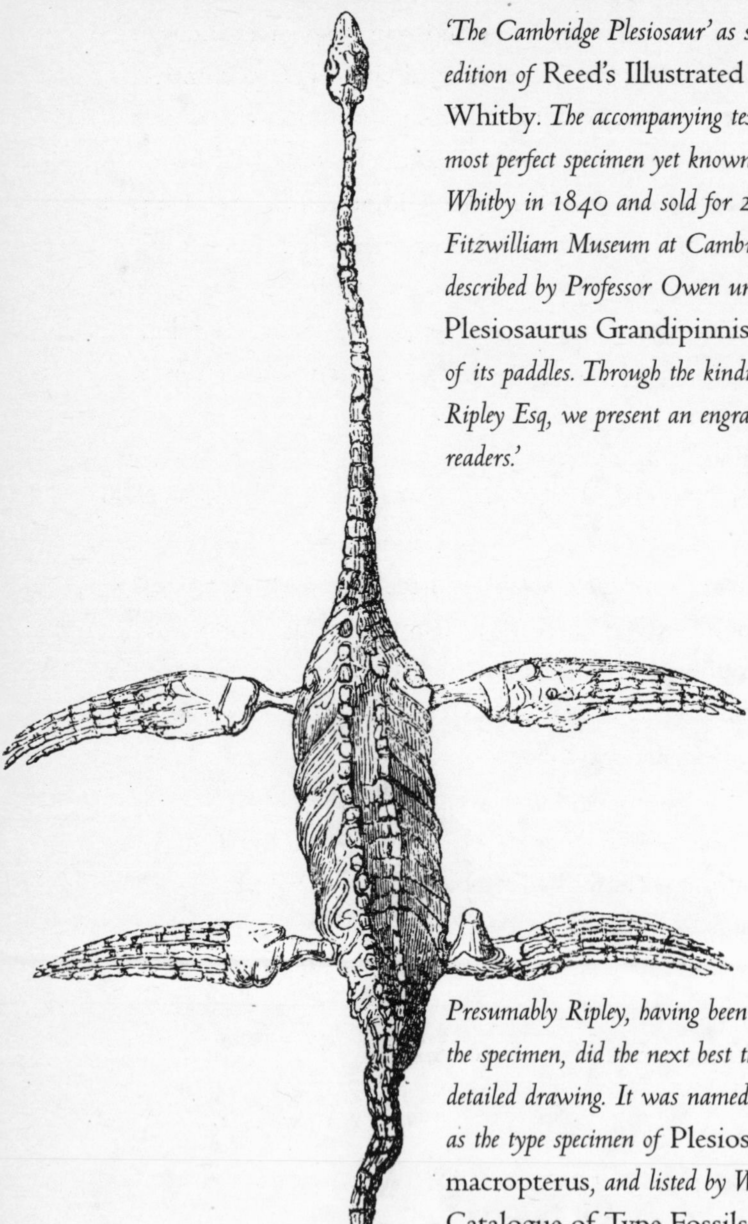

*'The Cambridge Plesiosaur' as shown in the 1854 edition of* Reed's Illustrated Guide To Whitby. *The accompanying text says, '. . . the most perfect specimen yet known was found near Whitby in 1840 and sold for 230l. to the Fitzwilliam Museum at Cambridge. It has been described by Professor Owen under the name* Plesiosaurus Grandipinnis *from the large size of its paddles. Through the kindness of Richard Ripley Esq, we present an engraving of it to our readers.'*

*Presumably Ripley, having been refused a cast of the specimen, did the next best thing and made a detailed drawing. It was named by Seeley in 1865 as the type specimen of* Plesiosaurus macropterus, *and listed by Woods in the 1891* Catalogue of Type Fossils in the Woodwardian Museum, *as* Eretmosaurus macropterus *(Seeley).*

names I am utterly astonished – in respect of their meanness. But there are several sits-off[?] of a more pleasing character. For instance Hopkins, who dined with me yesterday, mentioned incidentally that he intends to give £5. This is very liberal – like the man. A gratifying circumstance also occurred the other day whilst I was in your lecture-room. Col. Dealtry came in with Deck to visit the monster & was so much delighted that he begged to offer to Anstead his subscription of a sovereign.

I am very sorry to hear your bad account of your health. Your hand writing is less calligraphic than usual, which makes me fear the enemy has attacked your anterior members. However I feel satisfied that your Hypochondriacism & all peccant humours will vanish when you get your friends about you.

So I trust you will do this forthwith – & we wish you a very merry Xmas.

Nothing will give us all, Johnny & his dog included, greater pleasure than to see you at Norwich on some convenient occasion.

In the meantime live and be well My wife requests me to say that she has all the accts respecting the Plesiosaur in safe custody –

Yours my dear Sedgwick
always truly
J. W. Clark
Cambridge Decr 16. 1841"

Sedgwick's biographers, Clark and Hughes (1890), state that 'In a few weeks £264. 18s. 6d. was collected; a sum sufficient to discharge all expenses, and to leave a small balance in hand with which the Woodwardian Trustees agreed to purchase other fossils.'

The plesiosaur was displayed in the Woodwardian Museum in Cambridge, and Whitby Museum never did obtain a cast of the specimen. The fossil was moved to the new Sedgwick Museum of Geology in Cambridge when it opened in 1904. It remained in the same position until the early 1980s, when a new Holiday Inn was built nearby, on the other side of Downing Street. The construction work on the hotel caused vibrations which opened cracks in the fossil. The specimen was in any case in need of conservation work, so it was taken down from its mounting, cleaned, restored and placed on a horizontal bed of moulded fibreglass. The plesiosaur (ref no. J35182) has pride of place at one end of the museum named after the man who raised the money to buy it – and Whitby Museum did get its own plesiosaur, just four years later.

chapter fourteen

# Geology and Repose

## *The laws and operations of nature*

"There is, perhaps, no place in the world where a person may begin his geological pursuits with more advantage, or the experienced geologist revel with more pleasure and satisfaction, than along the Yorkshire coast. The strata rising up from the sea beach in succession, and stretching away for miles in the lofty and precipitous cliffs, give a strong and accurate impression of the manner in which substances have been deposited upon the earth's crust by the agency of water, and the manner in which they have afterwards been bent and fractured. The fragments of rock from the cliffs above, and the shores laid bare by the retreating tide, also afford the amplest and most perfect means of minutely investigating the remains of animals and plants, there imbedded, in a manner the most varied, interesting and instructive. And although Granite, and other unstratified rocks, exist in our district only as boulders, the basaltic dyke, which may be well examined a few miles from Whitby, gives no bad idea of rocks formed by the agency of fire.

In following these pursuits, which unfold the most expanded and fascinating views of creative power, and of the laws and operations of nature, the lofty hills towering above the spectator with their frowning summits, as in the neighbourhood of Whitby and

Scarbro', but more especially at Peak and Boulby, or corroded into fantastic forms of ruined castles and magnificent columns as at Flamborough Head, and the deep and solemn caves of the lias, hollowed out beneath by the raging billows of the ocean, leave on the mind the sublime sentiments of vastness, grandeur and awe; whilst in the interior of the district, the tabular and undulating hills with their fertile valleys, and meandering and silvery streams, glittering in the sun, afford scenes of varied beauty and interest; and the well-wooded ravines, sheltered from every blast, and joyous with the song of birds, invite to peaceful contemplation and repose."

Martin Simpson, 1855

chapter fifteen

# The Sixth Reptile

## *The Marquis and the museum*

In 1844 the annual British Association meeting (the principal gathering of British scientists) was held in York, hosted by the Yorkshire Philosophical Society. The Society was able to proudly display an enormous fossil plesiosaur, recently discovered in an alum quarry at Kettleness near Whitby. Both the meeting and the plesiosaur became centres of controversy.

William Buckland, one of the country's leading geologists, had published in 1836 a treatise on *Geology and Mineralogy Considered With Reference to Natural Theology*.*

The Dean of York, William Cockburn, had attacked Buckland's book when it was published. He now used the opportunity of the British Association meeting in his own city to renew the offensive. He spoke out against Buckland and the views of geologists, which were incompatible with his interpretation of the Bible. Buckland was not at the meeting, so Adam Sedgwick joined battle in his stead, and gave a full-blooded response on behalf of his science.

Unfortunately York Corporation had planned a celebratory dinner at the end of the conference, to which the Dean and Chapter of the Minster, together with all the leading scientists and the notable citizens of York were invited. The Lord Mayor of the time was George Hudson,

* This was one of the eight Bridgewater Treatises, commissioned in accordance with, and funded by, the bequest of the late Francis Henry Egerton, eighth Earl of Bridgewater. Each was to show the 'Power, Wisdom and Goodness of God, as manifest in the Creation'. Buckland's was by far the most successful and widely read.

the 'Railway King' who controlled the city like a medieval baron. It was a difficult situation for Hudson. The two opposing camps could hardly be accommodated at the same feast, though neither showed any signs of voluntarily withdrawing. Someone – either the scientific guests of the City, or its own churchmen – would have to be offended. In the end Hudson withdrew the invitation to the geologists, and in a notable phrase declared, 'We have decided for Moses and the Dean.'

This snub from the City might have been a minor irritation for the geologists, but for the Yorkshire Philosophical Society worse was to follow. The treasured plesiosaur, which they dared to hope would find a permanent home in their museum, was about to involve them in another dispute. The specimen was first briefly described by Edward Charlesworth, Curator to the Yorkshire Philosophical Society's Museum, in the Annual Report of the British Association for 1844.

"Notice of the Discovery of a large Specimen of *Plesiosaurus* found at Kettleness, on the Yorkshire Coast.

> The subject of this notice had been found a short time previously in the lias shale, quarried for the manufacture of alum, in the Kettleness Cliff, a few miles north of Whitby; and the lessees of the works, Messrs. Liddell and Gordon, had permitted the author to remove it, for the purpose of examination, to the museum of the Yorkshire Philosophical Society. Its total length was fifteen feet; that of the head above two feet; the neck, double that of the head; length of the humerus thirteen inches; length of femur, fourteen inches. The author observes, that the only published species exhibiting the above relative proportions of head and neck, is the *Plesiosaurus macrocephalus* of Conybeare, to which he supposes the present fossil must be referred.

To agree however fully with the characters assigned to this species by Prof Owen, the respective lengths of the femur and humerus should have been twelve and fourteen inches. He also finds the tail more depressed than it appears to have been in the celebrated specimen of *P. macrocephalus* belonging to the Earl of Enniskillen. The author in conclusion, regretted not having had time to make a more rigid examination of the Kettleness fossil, and stated his intention to publish a detailed account on some future occasion."

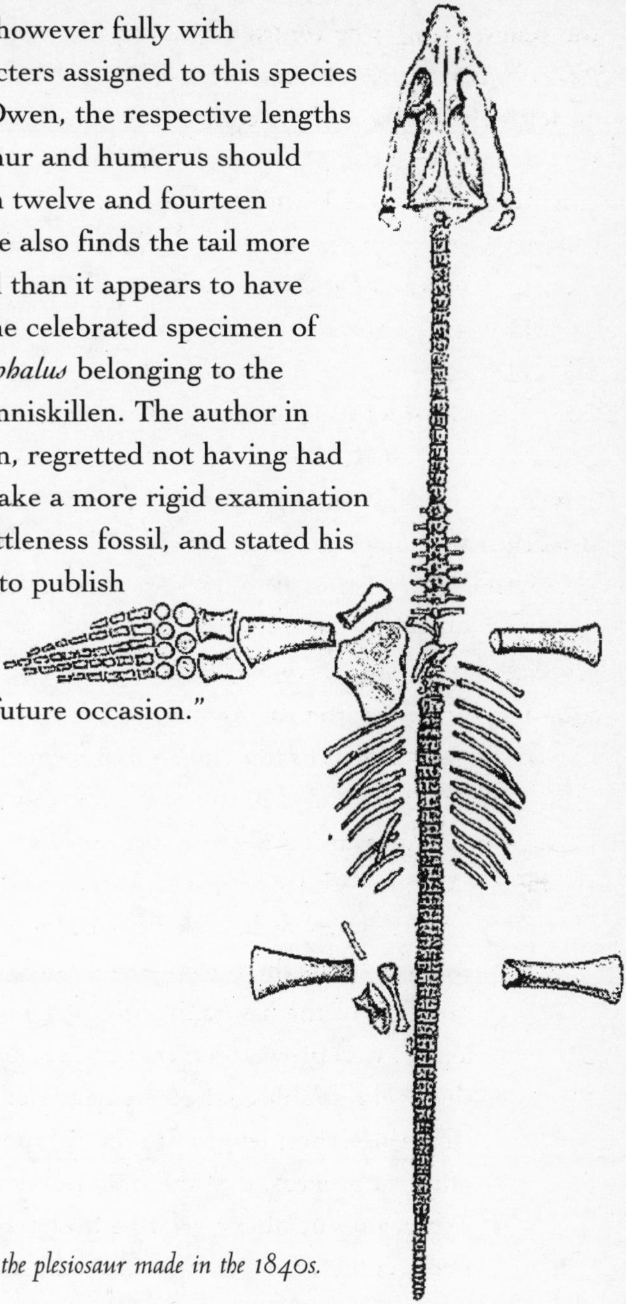

*Drawing of the plesiosaur made in the 1840s.*

Mr Liddell who, with his partner Mr Gordon, had leased the alum works at Kettleness from the land owner the Marquis of Normanby, had planned to donate the fossil skeleton to the Yorkshire Philosophical Society after the conference. But meanwhile the Marquis, who was incidentally an Honorary Member of the Whitby Literary and Philosophical Society, which had its own fossil collection, was taking a proprietary interest, as this exchange of letters shows. William Vizard was, we believe, Lord Normanby's agent.

"To The Marquis of Normanby
My Dear Lord

I send your Lordship a copy of a letter I had this morning about the fossil & of my answer to it.

The solicitors here say they can give no answer about the item I requested after as they had not seen it till now.

I have sent to my accountant to learn of him where he got it.

I am
Your Lordship's
most faithful
Servant
Wm Vizard"

These are the letters enclosed; first that from Joseph Gordon to Vizard:

5 Royal Terrace, Edinr.
Aug. 30: 1844

"Wm Vizard Esqr.
Mulgrave Alum Works

Dear Sir

I presume you have heard that in the course of their quarrying Messrs. Liddell and Gordon found a fossil Plesiosaurus in detached parts, which, by much care and at some expense they saved from the burning kiln and had collected and put together so as to exhibit a pretty perfect specimen. For its further improvement they have sent it to the Keeper of the York Museum – an accomplished Geologist, to have it better cleaned and set up correctly, in the view of having it exhibited at the Meeting of the British Association at York next month. My friends are advised that Lord Normanby as proprietor of the Rock has a legal right to claim the same interest in the Fossil as in any other product of the rock – a Lordship on its price or marketable value minus cost of production and sale – and I am desired by them to enquire thro' you how Lord Normanby would choose to have it disposed of for mutual gain, in case his Lordship be *not* disposed to concur with Messrs. Liddell & Gordon in presenting it to the York or some other public Museum.

I remain, dear Sir
faithfully Yours
Joseph Gordon

> Mr Liddell estimates the money value of the Fossil at £300 – or thereabouts. The cost of its acquisition he has not mentioned."

Mr Vizard felt able to reply on Lord Normanby's behalf, since they had obviously discussed the subject previously.

> Lincolns Inn Fields
> 4 Septr., 1844
>
> "John Gordon Esqr.
> Dear Sir
>
> Lord Normanby let to Messrs. Liddell & Gordon the Alum Rocks for the purpose of making Alum, and not to dig for Fossils and, I apprehend, they have as much to do with the Fossil you mention as they would have to do with a Silver Mine if they had found one and no more. Lord N will therefore enter into no treaty for the disposal of this article which he conceives belongs to himself alone.
>
> I am
> Your obedt. Sevt.
> Wm Vizard"

The outcome of the dispute was recorded in the Yorkshire Philosophical Society Council Minute Book, 8 March 1845:

> "Resolved that in consequence of the Arbitrators having decided that the specimen of Plesiosaurus deposited in the Yorkshire Museum is the property of the Marquis of Normanby a memorial be presented by the Council to that Nobleman."

At present, the memorial remains undiscovered, so we can only speculate on its contents. It is likely that the Society asked Lord Normanby if they could, in any case, keep the fossil. If so, they were to be disappointed. The letters referred to at the next council meeting have not been discovered, but the tone of their content is clear.

> "31 March 1845
> Letters have been received from Ld. Normanby respecting the Plesiosaurus reflecting upon the conduct of the Council in removing the specimen from Whitby as well as exhibiting the same at the late meeting of the British Association and Mr Charlesworth having written an explanation of the way in which he was permitted by Mr Liddle [sic] the lessee of the Alum Works, to remove the same, resolved that Mr Harcourt's offer to write to Ld. Normanby explaining the circumstances be gratefully accepted."

The fossil was duly taken back to Lord Normanby's home at Mulgrave Castle. It next appears in the annual report of the Whitby Literary and Philosophical Society for 1847, where no mention is made of its disputatious past.

> "Another most valuable acquisition has recently been obtained by purchase. It is a whole specimen of that rare fossil animal the Plesiosaurus, found at Kettleness in the summer of 1844. This magnificent fossil which was exhibited at York in 1844, and which has been since kept at Mulgrave Castle, differs materially from the specimen which went from Whitby to Cambridge in 1843 [actually 1841]. That was the Plesiosaurus

> Dolichodeirus, having a long slender neck, and a very small head; this is the Plesiosaurus Macrocephalus, having a stronger and shorter neck, and a large head, in which the bones and the teeth are distinctly seen; each measures about 15 feet. The Cambridge specimen cost about 220 guineas; the price of this is 200 pounds, but as 50 pounds is deducted as a subscription towards the purchase by the Marquis of Normanby, only 150 pounds more will require to be paid . . . The Friends of Science will not grudge to make the sacrifice required. When these specimens are duly arranged in our Museum, it will present a superb set of Saurian animals, such as perhaps no other Museum can boast . . .
>
> Many strangers have been brought here by the attractions of our museum, and when it is elevated to its high rank now in prospect, as possessing one of the finest sets of fossil organic remains of Saurian animals in the world, its attractions will be doubly powerful, and the advantages resulting to the Town and neighbourhood proportionally great."

A few years later, a guidebook to Whitby contemplated the appearance and behaviour of the plesiosaur when alive:

> "The Plesiosaurus, as its name implies, was more allied to the lizard than the fish, especially as regards its vertebrae. Like the Ichthyosaurus it united in its structure various types of organization. It had the head of a lizard, the teeth of a crocodile, a neck of enormous length like the body of a serpent, the back and tail of a quadruped, the ribs of a chameleon, and the paddles of

> a whale. It is supposed to have floated upon or near the surface of shallow waters, arching its long neck like a swan, and using it to dart at its prey."

So, after their intense disappointment over the loss to Cambridge four years earlier, the Whitby Museum got themselves a plesiosaur. Originally named *Plesiosaurus macrocephalus,* the specimen is now built into the wall of the new museum, where it is labelled *Plesiosaurus propinquus* (properly *Rhomaleosaurus propinquus,* see note on p. 258), specimen number WM 851S.

chapter sixteen

# The Strange Case of the Hyenas' Bones

## *William Buckland and the Kirkdale cave*

"Truth is not other-worldly, but is of this world. Wisdom does not lie in turning away from appearances but in mastering them."

Giorgio de Santillana

There is no plaque or monument to mark the Kirkdale cave. In local guidebooks it is mentioned as a curiosity, to be taken in at the end of a visit to nearby St Gregory's Minster. Even then the casual visitor might wonder if this could possibly be the right place. An unmarked track leads from the church across a footbridge over Hodge Beck and into a disused quarry – just like hundreds of other quarries dug into the oolitic limestone of this part of the North York Moors. The quarry walls are 30 feet high, and halfway up on the eastern side is a hole about three feet high and six feet across. The hole is horizontal, with a shelf projecting underneath. Anyone with reasonable agility can scramble up to it and peer inside. Even with a light you will not see very much – to be honest, this does not look like a very interesting cave. There is some litter on the floor near the entrance, and if you crawl back into the cave you will find that it gradually narrows, its limestone walls and roof closing in to form an uninviting tunnel. It is apparently innocuous. But the Kirkdale cave, and the man who made it famous, utterly changed the way we look at the history of life on Earth.

## DORSET

You will probably have come across as many who call themselves 'veterans' of Waterloo as would fill that battlefield three times over. And nearly as many poor wretches who would tell their tale of that day for the price of a glass of gin. So I will say only that the glorious victory meant nothing more to me than an aimless shambles played out amid clouds of smoke on fields of mud. I was wounded by French grapeshot late in the morning, when there was still enough organisation in the world for someone to throw me over a horse and send me back to a place of safety. I lost enough blood to fill my left boot, but was lucky not to go under the surgeon's (call it butcher's) knife. Most who survived their injuries did not survive their treatment.

I lingered on in the army for a while, becoming more aware of my uselessness within the military, and more afraid of life outside its cold embrace. In two years I resigned my commission, and walked out of the barracks, a free man, with little talent and no idea what to do with it. My health was not good, and precluded me from taking up any active occupation – at least for the time being. I had a small income from my late father's estate, and, though this was not enough to live comfortably in London, I decided to make the most of circumstances and took a cottage on the south coast, in a village in Dorset called Charmouth.

There was little enough to do in Charmouth. My housekeeper looked after my material wants, which left me time to do what I might. I had never been over-fond of reading, seeing myself, I suppose, as that oddly inadequate creature, a man of action. So I spent several hours each day walking among the chalk hills and along the coast. As my health improved, and the days lengthened, the walks got longer too. By the end of the summer I would think nothing of going ten miles past Lyme Regis (itself three miles from my house), lunching on a cliff top, and then returning in time for supper.

If I try to recount what my thoughts were during those hundreds of hours, I find myself disappointed. My mind, it seems, was mostly engaged in churning over the trivia of daily life, while occasionally lighting on those profound themes and events that were enough to bring my strides to a sudden halt, and to pull the horizon of my gaze to within an inch of my nose. So perhaps those hours were not wasted. Perhaps the churning contains some progression, so that after several hundred miles of solitary walking, I was able to withstand the shocks of remembrance a little more steadily, and to lift my gaze to the world around me.

I had never previously taken much interest in what you could call my natural surroundings. The sea was bracing, but the only point of interest on it was the occasional passing boat. The cliffs were suitable for walking, and were suitably deserted for my purposes (my ill-health had taken away my need of human company). The beach was also ideally suited for contemplation. And if I wanted the stimulation of watching human activity without the irritations of involvement, I would sit alongside the sea front at Charmouth or Lyme and watch the fishers go about their business. Land and seascapes had for me been something to cross. Nature was either the backdrop on which man played out his destiny, or a wilderness, useless and full of danger.

It so happened that Lyme Regis, at that time, was getting a reputation for a new pastime among the educated classes – the collecting of fossils. Interest had been stimulated by the singular activities of one family, the Annings, and in particular by the daughter, Mary. It was said that she had unearthed the entire skeleton of a reptile from the cliffs, and had sold it for more than £200. She now made a good living selling her finds from her mother's fish shop in the town. But money was not the main objective of those who collected for themselves. Indeed it is a little difficult to say what their, or should I say our, motive was. It perhaps began as a fashion, but lasted far too long for that. It was a combination of circumstances that first stimulated my own interest in the fossils of the Charmouth cliffs. I had grown quite bored by lack of mental exercise, and was beginning to be irritated with my own company. One morning, after a particularly heavy storm, I took a stroll along the beach near my house. The day was, I remember, quite still and the debris of the storm lay littered across it like the bunting from some wild carnival that had just left town. There were branches across the coast road and a tree had come down next to the churchyard. On the beach the sea lay like an innocent reveller, sighing in its drunken sleep, while all around was strewn the devastation of its night of mischief.

Every 50 yards or so a section had been torn out of the cliff and thrown down on to the beach. I walked over to the nearest pile of newly liberated stone, and turned a largish piece over with the toe of my boot. What I saw astonished me. The whole of the underside of the stone was encrusted with small coiled shells, each about two inches across. There were perhaps twenty or thirty packed tightly together, each complete, each perfect. It was one of the most beautiful things I had ever seen – the more so, I think, because it was so unexpected. I turned over a few other stones, and found some more of these ammonite fossils, but none was so strikingly presented as that first rock. I determined to have the fossils in my house, and began to carry the rock back up the beach. I had to pause every thirty paces or so – the piece must have weighed twenty pounds – but within an hour it was sitting on the little-used desk in my front parlour, while I picked pieces of sand and shingle out of its crevices.

A few weeks before, I might have felt a touch embarrassed at such behaviour – thinking it childlike and naive. But now a change had come over me. I had grown past that faux-worldliness that comes with adolescence, and infects some men for the rest of their lives. I had regained my enthusiasm for life, and with it my curiosity, of which this splendid fossil find was both cause and symptom. I was not a fossil-finder in the Mary Anning league, nor, to tell the truth, did I feel the urge to be. I found instead that the creatures themselves fascinated me. The very fact of their preservation, and the information they conveyed about some previous world, were worth more to me than any spectacular size or shape they might have.

A day or two later I received an invitation to dine at my neighbour's house. I had turned down several of these in the past, giving indisposition as a not altogether false excuse. This time I decided to accept. I will not trouble you with the conversations of Dorset dinner tables, except to say that the talk eventually turned, as in that part of the world it must, to geology. My host, Mr Dawson, a man in his fifties, was an amateur naturalist. Hearing of my modest find, he rushed off to his library and returned with an astonishing book entitled *The Organic Remains of a Former World* by John Parkinson. All polite conversation ceased as Dawson and I pored over the book – he as master and me very much as pupil. As the evening was getting late we arranged to meet on the beach the following morning at low tide.

Our conversation on the beach the next day, and over the next few days, was centred on the new ideas and discoveries that were being made by fossil collectors – some too wondrous to be believed. These were interrupted by the odd discovery of our own, though we never came across anything quite so beautiful as my first find. After a day or two I began to tell Dawson something of my life history up to that point. His open manner encouraged me to be frank with him about my sense of being somewhat adrift. He seemed to take a great interest in everything I said. In fact Dawson was one of those men who like to give the impression that they are slightly stupid, in order to be thoroughly engaging, while being extremely perceptive – and knowing that you will discover this in them eventually. I now see that he understood my situation perfectly, and was already planning how to effect the cure for it.

One midday, after an hour's walk along the coast, we took our rest on a low cliff top overlooking Lyme Bay.

As we both gazed out over the sea, Dawson said, 'If you are really interested in natural philosophy, and in the science of geology, then I might be able to help you.'

'I believe I am.'

'You might perhaps have heard of William Buckland?'

'I am afraid I have not.'

He turned to me and smiled. 'Then you soon will.'

It turned out that Dr Buckland was the son of Dawson's greatest friend, and had now risen to a senior post at the University of Oxford.

'He used to collect birds' eggs here when he was a boy. Got him started as a naturalist, you see.'

Dawson took obvious pride in his association with this man, and steadily persuaded me that I could do no better than to travel to Oxford to meet him. 'This is not the place for a man of your youth and intellectual energy. Why don't you go and see Buckland – I'll give you a letter for him – and get back in the flow of life. If you ask my opinion, geology is the coming thing. Get taught by Buckland and you'll be in demand from Dorset to Dundee.'

I took him up and left for Oxford within the month, carrying a letter of introduction and a head full of such geological knowledge as I could get from books. By good fortune I was to arrive in time to witness the greatest geological find of the decade, and one of the most ingenious pieces of scientific work that had ever been known.

## OXFORD

It would be unfair to call it a room when it was really more like a cavern, or a treasure house. High arched windows ran down either side allowing in light from the quadrangle and from the alley between the college and St Mark's church. But the light was soon obstructed and cast into strange arrangements of shape and shadow by the mass of objects that were piled in great heaps, as if they had been dropped from the ceiling. Mostly it looked like a stonemason's yard, with great blocks of differently coloured rock piled on the floor. There were hundreds of smaller stones too, and I could see that several of the pieces nearest to me were encrusted with fossils of different types. Among the rocks and stones there were odd bones lying about higgledy-piggledy, though each had a label tied to it with a neat piece of string. There was an occasional complete skeleton – I thought I recognised a dog and a small kangaroo. In addition there were glass cases containing stuffed animals, and everywhere piles of books, together with microscopes, hammers, scalpels and a whole assortment of other instruments. It was just possible to discern some furniture beneath the debris – a table here, a long bench there, and a settle, had all been pressed into service to accommodate the maximum amount of material. The way to proceed from the doorway was anyone's guess. I stepped over a piece of yellow sandstone the size of a sea-chest, only to find myself hemmed in by a specimen of coal decorated with the imprints of fern leaves, and a stack of what appeared to be the thigh bones of giants.

The room seemed to be unoccupied by human beings, but then as I negotiated my way past the coal I saw a hunched figure, half hidden by the skull of an elephant, sitting on a stool holding a large magnifying glass poised over a yellow bone. This, I knew, must be the famous Dr William Buckland. I studied him for a moment, as he remained totally absorbed and completely unaware of my presence.

I had been warned of his eccentricities, but my first sight of him was of the man of science, studiously going about his work. He wore a dark frock coat, which had become dusty with contact with his specimens; and his head, which was perhaps a little too large for his body, had a gnome-like appearance. His white hair, receding at the front, and long side whiskers, gave him the look of a country parson.

'Dr Buckland,' I called.

His head jerked up and he looked around, but didn't seem to see me. I stepped over the bones and peered round the skull. He drew back in mock horror.

'A live specimen of *Homo sapiens*.'

'I would like to introduce myself, Dr Buckland,' I began more nervously than I had expected. 'I have a letter for you from our mutual friend, Mr Dawson.'

'Dawson,' he said as if it was the name of a mollusc. 'How is he?' He took the letter and skimmed it across the room, where it landed neatly on a desk in the corner.

'Very well. He sends his best wishes to—'

'Good, now tell me – ' he handed me the magnifying glass, and pushed the piece of bone he had been examining along the bench towards me – 'what do you make of this?'

I took the instrument rather nervously, and first looked at the object without the glass. 'It is the bone of a mammal,' I said.

'Excellent!'

I thought the doctor's praise might be sardonic, so I mumbled, 'I suppose that is stating the obvious.' I had misjudged him.

'And nothing to be ashamed of in that. We must start from the obvious, from the known, and thence proceed to the conquest of the unknown, must we not?'

'Of course.'

'So, what else do you see?'

*Portrait of William Buckland by R. Ansdell, c. 1843.*

I picked up the bone, which was surprisingly weightless, and dark yellow in colour. 'It is not a fossil, though it seems to have been preserved in some way,' I said.

'Splendid. This is really first rate. Now, would you say that it is a complete bone?'

I looked at it carefully. I was conscious that I was yet to introduce myself, or to state the purpose of my visit, but the opportunity simply did not present itself.

'Not complete, no. It is broken, I would guess.'

Dr Buckland was becoming increasingly excited.

'Now, look through the glass at the ends of the bone. What do you see?'

'One end is the ball of a ball and socket.'

'Properly called a condyle.'

'The other must, I suppose, be broken.'

By now I was naturally intrigued – why was Dr Buckland so interested in what appeared to be a common or garden bone? I felt bold enough to enquire.

'Where did the bone come from, sir?'

'All in good time.' Buckland scurried off to another corner of his laboratory, from where he continued to address me. 'Your arrival is of great value' – he returned with a glass bottle, of the type used to hold acids and suchlike, I had seen them only in medical institutions – 'precisely because you know nothing of the circumstances of these bones. Your mind is therefore unclouded by prejudice. You must rely purely on observation and deduction, the purest form of natural philosophy.' He sang as he applied some drops from the bottle to the bone: 'Observation and deduction, observation and deduction.'

The bone failed to react to the reagent. 'Hmm,' said Buckland, a little disappointed, 'yet we learn even from what does not occur.'

I was about to make another attempt to state my name and

business, when there was a commotion in the corridor and the sounds of someone trying to push the door open.

'As for the origins of the bones,' Buckland said, 'I believe your questions are about to be answered.'

The door, we could dimly perceive, was now ajar, and a man's voice was heard: 'Dr Buckland, are you there?'

'I am indeed, Bishop. Do come in.'

We heard the shuffling of feet, and had the occasional glimpse of a grey head through the forest of artefacts.

'I hear you, Buckland, but how do I get to you!'

'Proceed along the north wall, Bishop, there is a way through between the skull of the plesiosaurus and the oyster bed from Kimmeridge.'

'I see, very good. Ah, I see you now.'

The face of the Bishop beamed at us as he emerged. Buckland stepped forward to help him over the last few articles.

'You seem to have accumulated a few more specimens since I was here last.'

'Ever more specimens, ever more knowledge of God's creations, Bishop.'

'Quite so,' said the Bishop, who then looked at me.

Buckland looked at me too, as if seeing me for the first time.

'Good evening,' said the Bishop.

'I'm sorry, My Lord, I haven't introduced you,' said Buckland. 'This is Dr Legge, the Bishop of Oxford, and this is . . .' Buckland looked worried. Did he know me?

'John Foster,' I said.

'Mr Foster,' Buckland agreed.

'I am very pleased to meet you, Mr Foster. Are you a fellow-geologist?' Dr Legge looked approvingly at the magnifying glass which I still held in my right hand.

'Oh, well, I suppose I am,' I blurted out, and then ploughed on, 'I have actually come to ask Dr Buckland if I might assist him in his work.'

'Splendid, Mr Foster.' The Bishop clapped my shoulder. 'Dr Buckland needs all the help he can get. Now Dr Buckland, I have something to tell you about these bones.'

'Good. Let us retire next door, where we can hear the whole story at your leisure.' Buckland led off with the Bishop following. I hesitated.

'You must come too, Foster. I have a feeling I shall have need of you.' He turned to the Bishop. 'Anything you tell me can be told to Mr Foster, Dr Legge, he has my complete confidence.' Then thinking this might be impertinent he added, 'I hope you don't mind?'

'Not at all, Dr Buckland, not at all.'

In this curious way I became, willy-nilly, an unpaid assistant to the great geologist.

Within a few minutes we were comfortably seated in a small sitting room with a red fire in the hearth.

'Now, Bishop, we are ready to hear your story; please do not omit any detail, no matter how trifling it may appear. Pray proceed.'

'This information has come to me from Mr Vernon Harcourt, so it can be relied upon, but it is not first hand.'

'I understand,' said Dr Buckland.

'It seems that in June of this year, a discovery was made near to the town of Kirkby Moorside in the North Riding of Yorkshire. A Mr John Harrison, a surgeon from that town, and Mr John Gibson, a visitor from London, were travelling along the newly made road, about a mile or two west of Kirkby Moorside, going towards the ancient ruin of Rievaulx Abbey, I believe. As they passed by a small hamlet, known as Kirkdale, famous only for the church of St Gregory, they noticed numerous bones of animals lying scattered on the new road.'

The Bishop leaned towards the fire, apparently to warm his hands, though it seemed to me that he was physically expressing his excitement at the tale he was unfolding before us.

'Having some interest in natural history, their curiosity was aroused. They made enquiries of the workmen, who were immediately able to tell them that the bones came from a nearby cave. As far as I, by which I mean Mr Harcourt, can tell, the mouth of the cave was uncovered by workmen in June, while quarrying stone for the road. As to how the bones got there—'

'One moment Dr Legge, if you please,' Buckland interrupted. 'Let us not begin theorising before we are in possession of the full facts. I beg your permission to ask a few questions about the cave.'

'Please do, Dr Buckland,' said the Bishop with boyish enthusiasm. 'I will answer whatever I can, though you understand I am only a second-hand witness in the case.'

'Thank you. It is important that you tell us only what you know, rather than what you surmise. Now, the bones – how many specimens are there?'

'Several hundred, I believe, perhaps thousands, of all shapes and sizes, and belonging to a whole menagerie of different—'

'Thank you, Bishop. I will attempt to judge the species for myself. The ones you have given to me are a representative sample?'

'I could not say that for sure, I believe they might be among the smaller bones. There are many different kinds. As I said in my note, Harcourt believes the complexities of the case to be beyond the competence of those amateur naturalists who are now at work on it. He requests your assistance.'

'And I am honoured by his request. The bones were found on the floor of the cave, I believe?'

'Indeed they were. Some had been taken by the workmen, who supposed them to be cattle bones, and thrown on the road. Many more have been taken by collectors.'

'Damn and blast!' Buckland jumped from his chair and walked around the room. 'Excuse my language, my dear friend, but are we to take it that the cave, this extraordinary site, that promises to teach us so much, has been stripped of its contents?'

'Not stripped, I understand, though reduced. But most of the bones are safely held in collections.'

'But not *in situ*!' Buckland seemed to have forgotten himself in his sudden anger.

'There may be some still *in situ*, Dr Buckland.' The Bishop flushed in response to the tone of Dr Buckland's question. 'But I could not say how many.'

Buckland regained his seat and a little of his composure.

'This amateur enthusiasm is very welcome. But we will never extend the limits of our knowledge if we continue to behave simply as collectors of mysterious objects. Pardon my frankness, Dr Legge, you are a valued friend, but these are serious matters.'

'So I see, Dr Buckland, and that is why all followers of natural history would wish you to involve yourself in this matter,' the Bishop replied with a calculated calmness.

Buckland paused a moment as if collecting his thoughts.

'Finally, to the cave itself. That has also been altered I believe. How large is the entrance?'

'It is now about three feet high by five feet across. The cave extends for some hundreds of feet into the hillside . . .'

'Which is of limestone?'

'Which is of limestone. The cave is generally low, though it enlarges in places.'

'Is there another entrance to the cave?'

'None has yet been found.'

'And the bones were found throughout the cave?'

'I believe they were.'

'Splendid, Dr Legge, thank you very much. Now, one last favour, if I may.'

'Of course.'

'Could you write to Dr Harcourt immediately and ask him to use whatever authority he possesses to prevent further material being taken from the cave, and if possible to prevent anyone entering the cave from now on.'

'I will write, but whether he will be able to—'

'Tell him, if you would, that he only needs to keep the cave closed for a further five days. Mr Foster and I will be leaving here on Wednesday, immediately after my morning lecture.'

'Then you will take it on?' the Bishop asked eagerly.

'Of course I will, Bishop. From what you have told me, and the samples you have kindly provided, it promises to be one of the most interesting problems I have ever encountered.'

## YORKSHIRE

'Let us review the evidence, Mr Foster.'

Dr Buckland and I were passing northwards through Woodstock in a carriage. I wish I could say that the journey was smooth, but the state of the roads made it barely tolerable. For the previous four days I had been bombarded with such information and argument about the science of geology and the politics of the venerable University of Oxford, as to make my head spin. By the time I attended his Wednesday lecture I was in no doubt that Buckland's great aims in his life were firmly to establish the study of natural history, and in particular geology, at the University; and to do this by demonstrating the astounding knowledge of God's creations that could be gained from the study of the rocks of the earth. If the reception afforded to his lectures is any measure (which he assures me it is not) then success is guaranteed.

On Wednesday morning the hall was packed with students from every college, though the natural sciences are optional to their courses of study. But these students had to fit as best they could in the back of the hall, as the seats at the front were occupied by the Masters of all the Oxford colleges and a good number of the Fellows and Members. During the lecture itself, in which Dr Buckland imparted information about recent discoveries of saurians in Germany and elsewhere, he kept his audience in peals of laughter by numerous humorous quotations, told in an exaggerated Devonian accent, and by his habit of dashing around the stage and audience, handing out specimens and imitating various exotic creatures. At one point he held the ends of his coat-tails out to each side and flapped vigorously, while leaping up and down, to demonstrate the flight of the pterodactyle. It was altogether a captivating performance – though in a sense it was not so much a performance as the extension of the man's personality. For Dr Buckland the excitement of his subject is self-evident, and is enhanced by the opportunity to share it with others – he is, I suppose, a man who loves an audience.

'Heavens above,' cried Buckland as we lurched in and out of a pothole, 'I've never understood why gentlemen have to sit in a carriage, while those of lesser status ride in comfort on horseback. Now, to the evidence.' He held tight to the arm of his seat and leaned forward.

'We have a cave do we not?'

'We do, Dr Buckland.'

'And we have some bones.'

'Several hundred.'

'And from our own examinations of the sample bones and teeth which Dr Legge procured for us, we deduce that at least five species of mammal, from the mouse to the mammoth, and three species of bird, are represented.'

'We do.'

'Good.' Dr Buckland leaned back in his seat, as the carriageway seemed a little more even.

'There is, too, a preponderance of hyena bones, Dr Buckland.'

'There is an *apparent* preponderance, Mr Foster. I should await our survey of a greater proportion of the cave contents before confirming that. Now, tell me, what do you find the most intriguing aspect of the information we have been given?'

'Well, I . . . to be honest, Dr Buckland, I find the whole thing intriguing.' I hardly knew what to say. I sensed that Dr Buckland had long ago decided on the answer to his question, and was testing my own reasoning abilities. I stumbled on.

'I mean, it is an amazing find, so many bones in one small cave, and of so many different species.'

'True, true, true.' The Doctor nodded sagely in time with his words. 'And we would be doing a service to the science of geology, and to human knowledge in general, were we simply to identify, and even reconstruct, those numerous species. But you see, Mr Foster' – he leaned across so that our faces were but two feet apart – 'we are going to do something much more than that.'

He suddenly burst into a cackle and threw himself back in his seat. 'Haha! Oh yes! We are going to do so much more than that.' I believe he would have done a jig if there had been room.

'You will have noticed, Mr Foster, that in England, the science of geology suffers from certain, shall we say, restraints that it does not have elsewhere – in Scotland and France and Germany for instance.'

'I regret that I am not familiar with the practice of science in those other countries, Dr Buckland.'

'Then let me tell you that in Scotland, and on the continent of Europe, geology is taught in institutions which are sympathetic to the growth of knowledge. They do not have to answer, as we do at Oxford, to the prejudices of a minority of churchmen.'

I must have shown surprise at this outburst, knowing Dr Buckland to be an ordained Anglican, as he hurried on: 'Don't misunderstand me, Mr Foster. It is not the Church of England which worries me; it is the limits imposed on knowledge by a minority of its higher officials. But enough of that, at least for the time being. We are, I believe, about to assist our subject in another way entirely. You are aware of the work of Baron Cuvier?'

'Indeed I am. They say he is able to construct a whole animal from a single bone.'

'You might be interested in this.' Buckland produced an opened letter from his pocket, which he gave me to read:

> "From Joseph Pentland, Esq., Paris
> I this morning received your kind letter of the 18th inst., and immediately communicated its contents to Mr Cuvier, who desires me to write to you in all haste in order to request you to procure for him if possible some of the fossil bones lately discovered in such abundance in Yorkshire, especially those of the Hyena, as he is now engaged in that part of his new work which treats of fossil Carnivores."

The name of Cuvier was like that of some kind of demi-god, producing miraculous works which were passed to us mortals from his faraway celestial laboratory. It had barely occurred to me that he existed as an ordinary living man. Now here was a letter from one of his assistants. My hand may have trembled a little as I passed it back to Dr Buckland.

'It is important to remember that Baron Cuvier is the greatest anatomist who ever lived. But,' Dr Buckland smiled as he replaced the letter in his breast pocket, 'he is not a geologist. Now, in England we happen to have few volcanic rocks, but a great many secondary rocks, many of which are fossiliferous. Thus, while we may leave our Scottish and French colleagues to consider the origins of the Highlands and the Alps, we in England have other fish to fry! Our unique contribution is our expertise in relating the fossils found preserved in rocks, to the geological history of our country, and of the world.'

There was no need for me to assist in this conversation. Dr Buckland was now entirely oblivious of my presence and was effectively lecturing to the world passing our carriage window.

'Up until now, the study of bones has been thought to lie within the province of surgeons – leaving us, the geologists, to tinker with shellfish and plant remains! But what do surgeons know of the history of the Earth? The Kirkdale cave, Mr Foster, will extend the science of geology over the history of the vertebrates.'

He gazed out of the window as his thoughts veered off again. 'But we will, I believe, do more than that with this case.'

We had cleared the town and were once more being bounced along what passes for a highway. Buckland's attention came back to me after a moment.

'So, to return to the cave. In cases such as these, which will extend the remit of our science, we must exercise our mental capacities in two ways. We must make a general survey of the scene, in order to decide what the key points and arguments are, what the crux of the matter is, and where it lies. Then' – at this point Dr Buckland began to throw his arms about in the confined space of the carriage, putting the safety of my nose in doubt – 'we must marshall all the evidence and the experience of the long nights we have spent in our laboratories and rush in' – he rose out of his seat and half dived towards me – 'upon these points of contradiction. Only then will we break through them, and find ourselves in possession of the citadel of truth!'

I gaped a little at this, giving Dr Buckland the chance to return to his seat and to a calmer frame of mind. He held up the index finger of his right hand, and spoke with solemnity. 'There is one aspect of this case which is at present entirely contradictory, and which therefore must prove to be the key to this story. I leave you to consider what that might be, while I attempt to find solace in sleep. Please wake me when we reach the red beds at Taschbrook.'

Before I had a chance to ask 'How will I know?' he had shut his eyes and was lost to the world.

It was a bright November day in 1821 when Dr Buckland first set eyes on the cave that was to make him famous. We had lodged at a comfortable inn at Kirkby Moorside – a market town of a few hundred souls, some twenty-five miles to the north of the great city of York. The country hereabouts is dominated by two features: the elevated region which forms the North York Moors, the Tabular Hills and the Cleveland Hills; and the extremely flat low-lying plain of the Vale of Pickering, which lies to the south of these hills and moors. The Howardian Hills, a southerly extension of the moors, run down the western edge of the Vale of Pickering, which is thought by many to be the remnant of an ancient lake. Kirkby Moorside lies at the foot of the southerly slope, where the moors come down to the vale. A series of streams cuts through the gentle escarpment, draining the hills and eventually falling into the River Derwent. The Derwent escapes from the Vale of Pickering not via the coast, where its route is blocked by a low range of hills, but by a narrow gorge at Kirkham in the west, 'the stoppage of which' (to quote Dr Buckland) 'would at once convert the whole of the Vale of Pickering into an immense inland lake'. As to

*The entrance to the Kirkdale cave, as depicted in Buckland's account of his studies,* Reliquiae Diluvianae.

the geology, the southern boundary of the moors is made of a belt of limestone which extends for thirty miles eastwards from the coast at Scarborough, to Helmsley. This rock dips gently to the south, and passes under the substrata of the Vale of Pickering, which are composed of stratified blue clay, which Dr Buckland declared identical to that at Oxford and Weymouth, where it also overlies oolitic limestone.

We travelled to the cave that morning in the company of the Reverend Eastmead, who volunteered to be our guide. We found it easiest to walk along the road from the town, bringing a pack-horse to carry Dr Buckland's pieces of equipment. About a mile from the town, as Dr Legge had described, we came to a ford over a small stream called Hodge Beck, but instead of crossing we turned north up a narrow track and almost immediately found ourselves in a limestone quarry. Directly facing us about eight feet above the floor of the quarry, was the cave entrance. The limestone of the quarry wall lay in regular horizontal strata, each between two and three feet thick. Every six feet or so the strata were cut into by vertical fissures or fractures, which may have been made by the percolations of groundwater. The effect was to give the limestone the appearance of a wall made of huge bricks, with the entrance to the cave as if one of the bricks were missing.

Although the wall looked, at first sight, to be vertical, it was easy enough to scramble up to the shelf that projected under the cave entrance.

'It was here,' Eastmead was saying, 'that the first bones were found *in situ*, Dr Buckland.'

'Yes,' said Buckland, looking at the floor at the entrance to the cave, 'and there are none left here now.'

'I'm afraid not. You see—'

'What I suggest, Dr Eastmead,' Buckland interrupted while peering into the entrance to the cave, 'is that Mr Foster and I go into the cave and make our investigations, and then afterwards we can ask you any questions that arise.'

'Quite so.'

Buckland and I returned to the pack-horse and brought a bundle of torches, magnifying glasses, hammers, knives, rulers, small hessian sample bags, candles, matches, knapsack and notebooks and, last of all, a bundle of string. Back at the mouth of the tunnel Buckland became serious. 'Dr Eastmead, you will please wait here while Mr Foster and I enter the cave. From the appearance of the floor, which is the most interesting part of the cave, it looks as if an entire herd of buffalo have stampeded through. We will therefore need to get to those parts of the cave which are least disturbed by our predecessors. I only wish I had been here before such damage was inflicted. But no matter. We will pay out this line of string, so that should the cave divide, you will know which way we have gone.' He took out his pocket watch. 'If we do not return within three hours, that is, by eleven-thirty, then please take steps to find us. But do not come on your own. While I am sure there is little danger, it is possible for a man to become wedged in these narrow passages. Now, Mr Foster . . .'

'Yes, Dr Buckland.'

'I will precede you, and will carry the torch. You will please ensure that the matches are easily to hand. When we reach a place of interest, we will stop and light two candles. I will collect samples and take notes and measurements. If I go on too far ahead, then call out. Perhaps you would be good enough to place the string in your knapsack and check it every so often. It is very important that we observe the floor of the cave, and must therefore tread only where the floor is already damaged – perhaps you will tread in my footprints.'

'Very good, Dr Buckland.'

As we stood in the cold sun arranging our equipment I, feeling excited and nervous, blurted out, 'Dr Buckland . . .'

'Yes, Mr Foster?'

'The one aspect of this case, as you mentioned in the carriage yesterday . . .'

'Yes?' Buckland raised his eyebrows and looked expectantly at me.

'. . . that is self-contradictory, and must therefore hold the key . . . '

'Yes?'

'Is it, by any chance, that all the animals whose bones are found here . . .'

'Yes?' Buckland was willing me to say the right words.

'. . . are found living today only in the tropics, while we are here in a temperate country, indeed an island . . .'

'Splendid! Excellent work, my dear Foster, excellent. Now, light this torch for me and let us proceed to our quarry. Oh dear, a frightful pun, forgive me.'

So we ducked our heads and half crouched, half crawled into the low mouth of the cave.

'Is that it, then, Dr Buckland, is that the key to the case?'

'It is, Mr Foster, an aspect that must be included in any solution, but it is not, in my view, the key. Never mind, let us observe! Observe and deduce! Observe and deduce!'

With that Dr Buckland dropped down on to his knees and edged forward, holding his lighted torch aloft.

It is remarkable how rapidly the absence of the sun's beneficial light transforms a place. We were no further than ten feet into the cave when a still, cold, echoing darkness wrapped itself around us and began to invade my very flesh and bones. It was enough to be going into such a place, where a man's senses might themselves conjure demons out of nothing. But to know that the cave contained the bones of so many creatures put an extra chill into my heart. Human imagination is a benign and wonderful instrument, but one that may be distorted against its possessor. It would not have seemed extraordinary to me that day to feel the hot breath of a hungry hyena upon my neck!

All I could see of my companion were the soles of his boots in the reflected light of his torch as he crawled ahead of me. After ten or fifteen feet of crawling, the light from the torch became more diffused and I sensed that we were entering a small chamber. We had room to sit up, and light candles.

Buckland took a magnifying glass, a measuring stick, a hammer, several sample bags and a notebook from his knapsack, and at once began an investigation of the chamber. His manner of executing this was so strange as to be almost comical. He immediately began scuttling about the cavern, sometimes stopping, occasionally kneeling and often lying flat on his face. He chattered away to himself under his breath the whole time, making notes in his book, and sometimes tapping away with his hammer and dropping a piece of material into one of his bags. He had become entirely unaware of my presence. I was put in mind of a bloodhound, quartering through a copse in search of a scent. Indeed Buckland was given to whimpers and moans that were extremely hound-like. After twenty minutes or so he declared himself satisfied, and we packed up and moved on, travelling ever further into the hillside.

Some bones had been left by the collectors, though the floor was damaged beyond its original condition in all but the farthest recesses. Dr Buckland was liberal in his complaints against the 'local philosophers'. In all there were three or four caverns where we were able to take some leisurely recordings, the remainder of the time we were on our hands and knees. It was with some relief that I heard Dr Buckland say, after about an hour, that he had seen enough, and that we should retreat.

Dr Eastmead was waiting outside.

'Well, Dr Buckland?' he said as we attempted to remove some of the dirt and dust from our clothes.

'Well, Dr Eastmead.'

'Have you found what you were looking for?'

'I have seen enough to convince me that my original conjectures about the cave were correct.'

'Then you have a theory about how the bones came to be here?'

'Indeed I do, Dr Eastmead, but this is not the time or place to expound it. I suggest we return to the town, so that I may catalogue my samples while my memory of them is fresh.'

'Dr Buckland, may I present the Reverend George Young of Whitby. Dr Young, Dr Buckland.'

We were sitting at an open log fire in the comfortable front parlour of the George and Dragon Hotel in Kirkby Moorside. Dr Buckland had refused to say more about his theories concerning the cave, and instead had spent the afternoon cleaning and arranging his own samples, and then viewing the extensive collection of the Reverend Eastmead.

We stood to welcome the newcomer, who had ridden twenty miles to meet Dr Buckland.

'I am honoured, sir, to make your acquaintance,' said Young as they shook hands vigorously. Though they were to enter into severe disagreement in the months ahead, there was no mistaking the real warmth of their mutual greetings – after all, they shared the same passion for geology, however different their views of the history of the Earth.

Dr Buckland introduced me as his helper; we ordered refreshments for the Reverend Young and, after the usual pleasantries, we resumed our seats.

Dr Buckland asked Young about the progress of his book on the 'Geology of the Yorkshire Coast', and asked that his name be added to the list of subscribers. The innkeeper brought us some glasses of winter ale and some meat and bread for Dr Young. The conversation moved on to the reason for our trip.

'You have inspected our glorious bone cave, Dr Buckland?'

Buckland ignored the presumptuous possessive pronoun. 'Indeed I have. It is a most interesting site.'

'I presume, Dr Buckland, that you agree with all other informed opinion on the origin of these bones?'

I had expected that Dr Buckland would choose this moment to propound his theories of the cave, but for a man who was normally so garrulous, he was uncharacteristically wary of speaking his mind. He gazed into the fire for a moment.

'Perhaps you will tell me, Dr Young, what this informed opinion is.' Buckland looked over at me and raised an eyebrow. At this invitation Dr Young pushed out his chest a little.

'Given that these are the bones of tropical animals, and we inhabit a northern country, an island moreover which any animal could not reach, it seems clear that these bones were washed from the tropics to the northern latitudes on the waters of the Great Deluge. As the waters rose, or subsided, some bones drifted into the Kirkdale cave, perhaps due to the peculiarities of the currents, where they have lain undisturbed ever since.' Dr Young leaned forward and waved a crust of bread decisively in the direction of Dr Buckland. 'I had come to this view before I arrived at Kirkdale, but my own inspection of the cave, with its deposit of muddy sediment, confirmed me in my opinion – which is now held by everyone who has seen the evidence.'

'Then you are to be congratulated, Dr Young.'

Young flushed a little. 'Thank you for your kind words. I take it, then, that you are in full concurrence with this interpretation of the events?'

Buckland was ready to avoid answering. 'There are a few facts which lead me to hesitate from such a conclusion.'

'Oh really? What are those facts?'

'I cannot go into more detail at the moment; you will appreciate that it is only twelve hours since we first entered the cave.'

'Of course. But I cannot see how any other explanation—'

'Tell me, Dr Young' – Buckland had a habit of interrupting people when he did not like what he knew they were going to say – 'what is your own belief concerning the formation of the secondary rocks of the Earth – I mean particularly the time when this formation occurred.'

'This occurred during the Great Deluge. It is clear from the Book of Genesis that this was the only time when these rocks could have been formed.'

'And so the rocks which form the limestone walls, floor and roof of the cave were made during the Noachian Deluge.'

'Indeed.'

'And the bones were washed into the cave during the same deluge?'

'The rocks which surround the cave were presumably made earlier on in this tempestuous episode; the bones came later.' Dr Young tried his best to look untroubled by the absurdity of this proposition.

'I see.' said Buckland.

'It is the only explanation, is it not, since the whole history of Creation is given in the Book of Genesis. The primary rocks were made "In the Beginning" so to speak, while the secondary rocks, which contain the remains of some living creatures, were formed in the Deluge. The Bible will, I believe, allow no other explanation.'

## OXFORD

'Gentlemen, gentlemen –' Dr Buckland raised his arms to quiet the hubbub in the room – 'thank you all for attending this afternoon. I know you must be wondering just why you are here, while at the same time being aware of my recent journey to Yorkshire.' We were crowded into Dr Buckland's small office-cum-sitting room at Christ Church, the dozen or so men, who occupied every available seat, now quiet and intent upon the speaker.

'I have invited you to hear the discoveries which I, together with Mr Foster, and others, have made at Kirkdale. But more than that, I wish to demonstrate the reasoning which has led me to what many will call a quite extraordinary conclusion. I have deliberately invited you, gentlemen, as men of Oxford who are yet in possession of reason and of open minds.' There were some polite murmurs of laughter at this point.

'Let me start by saying that the facts presented at the Kirkdale cave afford one of the most complete and satisfactory chains of circumstantial evidence I have ever met with in the course of my geological investigations.' Utter silence fell, and the audience was once again enraptured. They were also fascinated, I believe, to see Buckland, normally so humorous, in such serious mood (though the theatricality of his method of delivery cannot be denied).

'Mr Foster will read from my notes on the subject. I will extemporise as necessary, and will stop for questions at suitable places. I beg you, gentlemen, to pay the utmost attention to the details of the case. Please, Mr Foster, proceed.' Dr Buckland sat down at the front of the room, while I, sitting to one side, began to read from his prepared notes.

After a brief description of the general location and geology of the Kirkdale cave, I moved on, via Dr Buckland's notes, to the cave itself.

'The entrance to the cave, which has been enlarged by quarrying, is no more than three feet high by five feet broad. The cave expands and contracts itself regularly from two to seven feet in breadth and two to fourteen feet in height, diminishing, however, as it proceeds into the interior of the hill. The cave is about twenty feet below the level of the incumbent field, the surface of which is nearly horizontal, and parallel to the stratification of the limestone, and to the bottom of the cave. Its main direction is east-south-east with zigzags to right and left, and is thought to be 245 feet in length. There are but two or three places where it is possible to stand up, where fissures intersect the cave. These continue upwards a few feet, and are then closed off. There is no access to the cave from the hillside above, the aforementioned being the only entrance.'

'Mark that well, gentlemen,' Buckland interjected: '"the only entrance", and note its dimensions. Pardon me, Mr Foster, please continue.'

I found my place in the text and read on: 'The sides and roof of the cave are studded with chert and stalactite, and the floor is covered in a bed of mud or loamy sediment.'

Dr Buckland, seemingly unable to contain the tension in his body, stood up at this point and paced around his chair.

'A moment if you will, Mr Foster. There we have, gentlemen, a description of a perfectly ordinary limestone cave. Now we will proceed to a description of the floor of the cave and its contents – which are very far from ordinary. Please, Mr Foster . . .' Dr Buckland resumed his seat with some difficulty and sat intent on hearing his own words being read to him. I turned a page and proceeded.

'The depth of the loam at the cave entrance is about a foot. There is no mud adhering to the sides or roof or fissures, nor anything to suggest it entered the cave through fissures. The sediment is flat, argillaceous and micaceous, as would be easily suspended in water, mixed with calcareous matter that may have come from the stalactite. The stalactite itself is like that in other caves; but on tracing it downwards to the surface of the mud, it was there found to turn off at right angles from the sides of the cave, and form above the mud a plate or crust, shooting across like ice on the surface of water, or cream on a pan of milk. A great portion of this crust had been destroyed by digging up the mud to extract the bones before my arrival; it still remained in some few places projecting from the sides, and in one or two, where it was very thick, it formed a continuous bridge over the mud, entirely across from one side to the other. There was nowhere any black earth or admixture of animal matter, except an infinity of extremely minute particles of undecomposed bone. There is no alternation of mud with stalactite.'

'Let me remind you, ' Dr Buckland addressed his audience, 'that the elevation of the cave, nearly eighty feet above the Hodge Beck, excludes the possibility of our attributing the muddy sediment it contains to any land flood or extraordinary rise of the waters of this or any other river in the neighbourhood. Also note that the lack of alternation of mud with stalactite indicates a single, extraordinary event.'

There were murmurs of assent, as it must have been obvious to most of the audience that this single event was the Great Deluge, as described in the Book of Genesis.

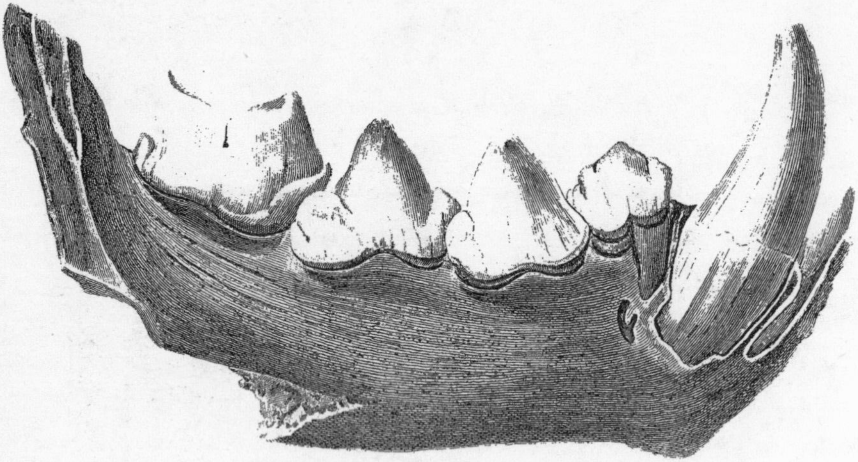

Above: *Jaw bone of hyena from the Kirkdale cave.*

Opposite: *Bone fragments of large mammals gathered from the Kirkdale cave, as described by Buckland in his book* Reliquiae Diluvianae*:* Left to right, top to bottom: *Small molar tooth of a very young elephant; Fragment of a still younger elephant's tooth; Molar tooth of upper jaw of rhinoceros; Inside view of molar tooth of lower jaw of rhinoceros; Crown of same as seen from above; Outside view of same; Molar tooth of upper jaw of a horse; Molar tooth of hippopotamus, not yet worn down; Different view of same; Molar tooth of hippopotamus, having the summits of the crown worn down.*

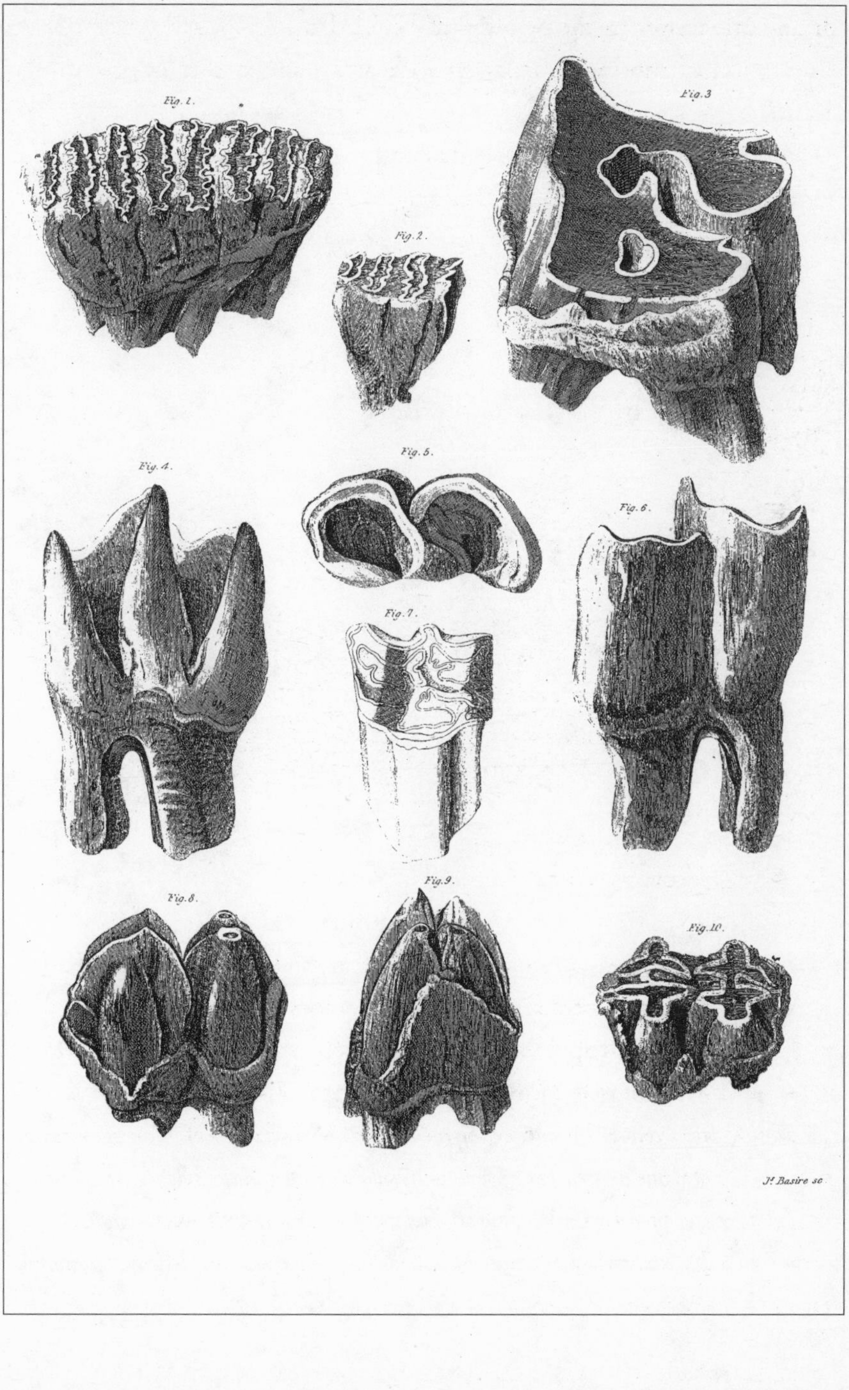
Fig. 1.
Fig. 2.
Fig. 3
Fig. 4.
Fig. 5.
Fig. 6.
Fig. 7.
Fig. 8.
Fig. 9.
Fig. 10.
Jˢ Basire sc

At a nod from Dr Buckland I read on.

'The bottom of the cave, on first removing the mud, was found to be strewed all over like a dog-kennel, from one end to the other, with hundreds of teeth and bones, or rather broken and splintered fragments of bones, of all the animals below enumerated; they were found in greatest quantity near its mouth, simply because its area in this part was most capacious; those of the larger animals, elephant, rhinoceros, &c. were found co-extensively with all the rest, even in the inmost and smallest recesses.'

'The list of species, Mr Foster, is at the back. Perhaps you could read it now.'

'Certainly, Dr Buckland.' I located a loose leaf in the back of the notebook. 'The species so far identified are: hyena, tiger, bear, wolf, fox, weasel, elephant, rhinoceros, hippopotamus . . .' There were sounds of growing incredulity from the far corners of the room as I proceeded, 'mastodon, horse, ox, deer, rabbit or hare, water-rat, mouse, some species of bird.' Dr Buckland held up his hand to postpone questioning on this incredible list.

'A moment more of your patience if you please, gentlemen. Go on with the text, Mr Foster.'

I divined that for those hearing of this for the first time, there was much that must seem fantastical. I returned to the text.

'The bones were in the mud beneath the crust, though where the mud was thin, the upper ends of some bones projected through the stalagmite like the legs of pigeons through a pie crust into the void above, and have become thinly covered in stalagmitic drippings. The bones, as with others found in caves, are not mineralised, but simply in the state of grave bones, more or less decayed. They have no connection with the rocks of the cave. The effect of the loam and stalagmite in preserving the bones from decomposition, by protecting them from all access of atmospheric air, has been very remarkable. In some bones, the whole of their gelatine has been preserved. The bones have passed through different stages and gradations of decomposition according to the different length of time they had lain exposed in the bottom of the den, before the muddy sediment entered, which, since its introduction, has preserved them from farther decay.

'Scarcely a single bone has escaped fracture, with the exception of the astragalus, and other hard and solid bones of the tarsus and carpus joints, and those of the feet. The hyena bones have been broken and apparently gnawed equally with those of the other animals. Not one skull is to be found entire; and it is so rare to find a large bone of any kind that has not been more or less broken, that there is no hope of obtaining materials for the construction of a single limb, and still less of an entire skeleton. However, I cannot calculate the total number of hyenas of which there is evidence, at less than two or three hundred.'

At this I heard several gasps and 'Upon my soul!' from near the window.

'The only remains that have been found of the tiger species are two large canine teeth, each four inches in length, and a few molar teeth, one of which is in my possession. These exceed in size that of the largest lion or Bengal tiger. There is only one tusk of a bear, *Ursus spelaeus*.'

I came to the end of a page and Dr Buckland laid a hand upon my arm. I waited while he stood and once again circumnavigated his chair.

'An extraordinary collection of animals, gentlemen, but please do not be distracted by the exotic nature of the species. Keep your attention on the state of the bones – every single one of them splintered and broken. Now before we apply our faculties of reason to this collection of evidence, we will hear just one more aspect of the case which will guide our thinking.'

'May we hope, Dr Buckland,' said a familiar voice from the

back, 'that you will be able to explain to us how a hippopotamus could enter a cave a mere two feet in height!' There was general laughter and a shout of 'On his hands and knees' before Buckland restored order.

'I congratulate you, Dr Legge, on going straight to the heart of the matter. Another minute, and all will become clear. Mr Foster . . .'

I once again returned to the text.

'My final discovery was of many small balls of solid calcareous matter. Its external form is that of a sphere, irregularly compressed, as in the faeces of a sheep; its colour is yellowish white. That concludes the evidence in the case.' I closed the notebook in as theatrical a manner as possible and waited. Dr Buckland allowed a moment's silence, as if he too was in awe of the evidence he had presented, and the conclusions to which it inevitably led. (Though one must also consider his weakness for dramatic gestures.)

He rose slowly from his seat.

'So there, gentlemen, you have it. A long low cave with a small entrance. Its floor carpeted in mud, which contains the broken and splintered fragments of thousands of bones of large and small mammals, and which is covered in a layer of stalagmite. The bones, incidentally, worn and polished on one side, sharp and rough on the other. And in addition small balls of a mysterious matter, resembling animal faeces in appearance.' One or two people coughed or adjusted themselves in their seats. They had grasped the information and were aware that Buckland was coming now to the focus of the evening – what did it all mean?

'Having studied the evidence long and hard, it is my considered opinion that there are but three theories which might account for the evidence.' There was a pause as the speaker and his audience collected themselves for the denouement.

'Firstly,' Dr Buckland held his right arm aloft with one finger raised, 'the animals had entered the cave to die, or to escape from some great convulsion. Against this we must argue the diameter of the cave, compared with the bulk of the elephant and rhinoceros. As to the smaller animals, we can imagine no circumstances that would collect together, spontaneously, animals of such different habits as tigers, bears, wolves, foxes, horses, oxen, deer, rabbits, water-rats, mice, weasels and birds. In addition, this circumstance does not account for the condition of the bones and the lack of complete skeletons.

'Secondly – ' Buckland raised another finger of his right hand: 'the bones were drifted in by the waters of a flood. This is the theory expounded by those who call themselves Scriptural Geologists. But again the bones would have to be in pieces and separated from the flesh beforehand, or else they would not fit through the cave entrance. Also they would be mixed with gravel and at least slightly rolled on their passage, yet in the interior of the cave I could not find a single rolled pebble, nor have I seen in all the collections that have been taken from it one bone, or fragment of bone, that bears the slightest mark of having been rolled by the action of water. Moreover it would remain to be explained how they came to be broken and splintered.

'The third explanation, gentlemen' – Buckland dropped his hand, slowed his voice and took his time to look at each and every one of the men in the room as he spoke – 'and the conclusion that we, as reasonable men, are irresistibly drawn to, is that the cave was nothing more or less than a hyenas' den. The bones that we have found were dragged in for food by generations of hyenas. All the animals whose bones are found in the cave, therefore, lived and died not far from the spot where their remains are found. They are, gentlemen, *the wreck of the hyenas' larder!*'

A moment's silence followed this announcement, and then it seemed a dozen voices were talking all at once – some to each other, some calling to Buckland, some simply expressing amazement to themselves. Buckland stood in the middle of the hubbub with an enormous smile on his face. Several gentlemen came over to him and clapped him on the back. He had his audience, and he had shown the fatal weakness of his rivals' theories. Now he must prove his own.

## THE TRAVELLING SHOW

'Thank you for your extraordinary patience, gentlemen, which is about to be rewarded in full.'

Dr Buckland turned to face the same dozen men as we gathered in a rough semicircle around him. We had walked in crocodile fashion with Buckland at our head, out of Christ Church, around the Cathedral and out to the Meadows at the back. A short tramp across the northern end of the field brought us to an encampment of tents and wagons, surrounded by a temporary picket fence. We had walked a few yards round to the entrance where a large gaudy sign was erected over a precarious wooden archway:

WOMBWELL'S TRAVELLING SHOW

*Animals, Serpents and Lizards from Africa, India and South America*

There were some crude paintings of various beasts on boards attached to the fence near the entrance, where a small queue was waiting to pay their pennies to enter. Buckland waved to someone inside the showground, who greeted him effusively and ushered us all through the barrier, whence we were led to the far corner of a group of wagons.

We were now standing in front of an iron cage, with a cart-wheel at each corner. There were a few curious members of the public gazing into other similar cages, from which arose an unpleasant stench. 'Now, gentlemen, observe please the condition of this bone, which was lately collected from the cave at Kirkdale.' We passed a large fragment of bone among us. It was about eight inches long and perhaps four in circumference. Both ends were broken, one in particular showing remarkably jagged and splintered edges.

'It has been identified as the shin bone of an ox. Now . . .' Dr Buckland waved to a tall man who had come to stand behind us, and who now stepped forward. He was theatrically dressed in high boots, riding breeches, a top hat and a red waistcoat, being obviously connected with the 'show'. He carried a hessian sack in his left hand.

'May I present Mr Wombwell, gentlemen' – Buckland shook the man's hand as he spoke – 'who has kindly assented to us carrying out an experiment using his precious animal.' Wombwell nodded a greeting to us and passed the sack to Buckland, who withdrew from it a freshly-butchered portion of an animal's leg. There was an immediate rushing sound behind him, as one of the ugliest and most unpleasant creatures I have ever seen emerged from a pile of straw in the cage, and began to pay keen attention to the piece of flesh. We all drew back a little, but Buckland remained unruffled. 'This is the shin of an ox, purchased by Mr Wombwell this morning in Oxford, and this,' he turned to the creature, 'is a Cape Hyena, is it not, Mr Wombwell?'

'It is, sir. From the south of Africa.'

'Now, gentlemen, Mr Wombwell will feed this shin bone to the hyena. Please pay close attention.'

The last request was entirely unnecessary. We watched transfixed as Wombwell, with complete sang-froid, passed the leg between the bars of the cage. The animal demolished the bone as if it were made of nothing stronger than bread. It began at the upper end, snapping off fragments of bone, and swallowing them whole as fast as they were broken off. The middle portion of the bone splintered into angular fragments under the tremendous pressure of its jaws. Many of these fragments were greedily consumed, while others had the marrow licked out of them. The lowest portion, too, was licked out for marrow, but the condyle, which is very hard and contains no marrow, was left. The power of the animal's jaws exceeded anything of the kind I, or I believe any of those present, had previously seen. It reminded me of nothing so much as a miners' crushing mill, or the scissars with which they cut off bars of iron and copper in the metal foundries. Within a couple of moments the bone was reduced to a few fragments.

Mr Wombwell then summoned one of his assistants, who pushed a spar of wood into the far side of the cage. While the animal was engaged in reducing this timber to splinters, Wombwell reached into the cage and retrieved several fragments of the ox bone, including the condyle. He solemnly wiped these on his handkerchief and passed them to Dr Buckland.

'Take a moment to compare these fragments with those we have seen from Kirkdale, and particularly the shin bone we saw not a moment ago.' Buckland passed pieces of the bone to Dr Legge and others. The similarities were, as everyone remarked, undeniable.

'In addition, gentlemen, please inspect these balls of matter' – Buckland retrieved some more material from the sack – 'which Mr Wombwell this morning recovered from the creature's cage, and these found in Kirkdale.'

As we inspected the bones, and the hyena played with the remains of the wooden spar, Dr Buckland warmed to his theme.

'To animals, such as the specimen before us,' he gestured to the malevolent beast, 'our cave at Kirkdale would afford a most convenient habitation; and the circumstances we find developed in it are entirely consistent with the habits of the modern hyena. When we ask, why has not a single entire skeleton of any animal survived in the Kirkdale cave?, the answer lies before us, in the power and known habit of hyenas to devour the bones of their prey. It is also recorded by Mr Brown in his journey to Darfur "that upon one of them", i.e. a hyena, "being wounded, his companions instantly tear him to pieces and devour him".

It seems therefore in the highest degree probable that the mangled relics of hundreds of hyenas that lie indiscriminately scattered and equally broken with the bones of other animals in the cave of Kirkdale, were reduced to this state by the agency of the surviving individuals of their own species.'

'But Dr Buckland,' a voice rose at the back of our group, 'is it possible that hyenas would have killed such animals as bears and tigers, let alone the elephant?'

'I agree that this is unlikely, Dr Kidd. It is more probable that the hyenas found the dead carcasses of the bear and tiger, and dragged them to the den, than that they were ever joint tenants of the same cavern. It is however obvious that they were all contemporaneous inhabitants of antediluvian Yorkshire. The bones and teeth of the elephant, rhinoceros and hippopotamus are, it seems to me, the remains of individuals that died a natural death. While a hyena would have the strength neither to kill a living animal of that size, nor to drag home an entire carcass, yet he could carry away piecemeal the fragments of the most bulky animals that died in the course of nature, and thus introduce them to the inmost recesses of the den.'

At this the hyena, a curiously made creature whose large head and shoulders seem out of proportion to its slender hindquarters, squatted and then lay with a deep self-satisfied sigh, on its straw bedding.

'Now' – Dr Buckland reclaimed the samples of grey faeces-like matter that had been found both in the cave and in the cage in front of which we stood – 'what do you make of this, gentlemen?'

'It is shaped like faeces, but its state of preservation presumably means that it is something else entirely,' said the man I now knew as Dr Kidd.

'I am so utterly pleased that you have all taken the time to come here, since your reasoning on this matter is so close to my own –

that is a great comfort to me. I confess I was ignorant of the importance of this substance, and was quite unprepared for it to prove to be the confirmation of my opinions. It is the keystone of the whole case. Dr Wollaston recognised this as a substance called *album graecum*, being the excreta of bone-eating animals. He analysed its contents and found that it contains exactly the same minerals as the bones themselves. It is, as you rightly say, Dr Kidd, a form of faeces, but one containing no decayed organic matter. It is therefore perfectly preserved.' Buckland held up the glorious piece of hyena faeces, that we all might reflect on its importance.

While we were doing so Buckland took his story a little further. 'The remains of many of the creatures from the Kirkdale cave have been found in the diluvial gravel formed by the waters of the great Mosaic flood, while their cousins live on in tropical regions. Indeed similar remains have been found in diluvial gravel beds all over northern latitudes. It was previously thought possible that they might have drifted north from equatorial regions, but the facts developed in this charnel-house of the antediluvian forests of Yorkshire demonstrate that there was a long succession of years in which the elephant, rhinoceros and hippopotamus, tiger and hyena, inhabited England and other regions of the northern hemisphere, in Europe, Siberia and North America. We have a theory, gentlemen, which embraces, not only Kirkdale, or even Yorkshire or England, but the whole of the northern world.'

It was a triumphant note on which to end, but as we were preparing to leave Dr Legge said, 'To return to an earlier question, Dr Buckland, how is it that these tropical animals could live in British conditions?'

'It is curious to observe, Dr Legge, but let me say that it is not essential to the point before me to find a solution. On some things we can be certain, and on others we may speculate – but errors in our speculations must not overturn our solid deductions. Monsieur Cuvier's opinion is that some of these animals were adapted to the rigours of a northern winter. This has support from the elephant's carcass found in Tungusia in Siberia, and the rhinoceros from the same frozen country, both preserved with portions of skin covered with long hair and wool. There are indeed species of the fox tribe which are adapted to both polar and tropical climates. This may be the solution, but it may equally be true that the climate of our lands has changed. At present we simply cannot say.'

'And the flood you refer to, Dr Buckland,' continued Dr Legge, 'this is presumably the biblical Deluge.'

'Indeed it is, Bishop,' said Buckland. 'Every detail of my theory is in full agreement with the words of the Book of Genesis.'

A year later, Buckland's great work on the Kirkdale cave, *Reliquiae Diluvianae,* was published. His theory of the hyenas' den caused a sensation that made him the most famous scientist in England – a man whose name could be spoken in the same breath as Cuvier.

A few months after that, with Dr Buckland's encouragement, I left Oxford on an extended visit to geological sites in France, Italy and, latterly, America. Buckland has, God bless him, continued to be a controversial, and often misunderstood, figure. I can still hear his voice singing out 'Observe and deduce, observe and deduce.' And I know too that he did not always follow his own motto – his theory of the Kirkdale cave was formed in his mind before he laid eyes on the place.

But this only enhances my respect for Buckland, as I have come to see that science is an argument of ideas more than an exchange of information. Beneath his eccentric and theatrical nature, Buckland really did create a revolution in science by thinking differently from anyone who had gone before him. And the Kirkdale cave gave him the opportunity to demonstrate his ideas to the world. Where Cuvier would have been able to identify every animal from the bones that remained, Buckland did something else entirely – as he promised he would. He used the evidence of the cave to re-create the world in which the Kirkdale cave and its animals existed. And what was truly revolutionary was that he used the study of the behaviour of present-day animals to do it. The visit to Mr Wombwell's show was pure theatrical Buckland, but it was central to his scientific genius. He understood that to know the past you must know the present.

In the 1820s the serious study of fossils was restricted to attempts at identification and classification. Within a few years of Buckland's work on the Kirkdale cave, the term 'palaeontology' was coined to describe the field of study that Buckland had opened, and the science of the 'study of ancient life' was officially born. In every field of human endeavour, each great innovation appears in retrospect to be a natural progression. Buckland may have been forced into reconstructing the lives of the Kirkdale cave animals in order to overcome Young's mistaken theory. But his ideas *were* shocking, exhilarating and controversial, and yet they now seem obvious – surely the truest sign of genius.

chapter seventeen

# The Seventh Reptile

## *Kettleness to Dublin*

The most magnificent of all the fossil reptile skeletons discovered on the Yorkshire coast was found on 27 July 1848 in the Kettleness alum quarry owned by the Marquis of Normanby. Unlike the previous find (see p. 203) this specimen was immediately removed to the Marquis's home at Mulgrave Castle. One of the flippers was destroyed in the alum crusher before the fossil was spotted. Apart from that one limb the plesiosaur skeleton, measuring 23 feet 4 inches long, and 13 feet across, was intact. It was on display at Mulgrave Castle for five years.

John Phillips wrote in 1853: 'Of four large specimens of Plesiosaur hitherto discovered in the lias of the Yorkshire coast, one (P. grandipennis of Owen) is at the Cambridge University Museum. Another (P. brachyspondylus of Owen) is at the Whitby Museum, a third is now at York; the fourth and most perfect of all, remains at Mulgrave Castle.'

Later that year the Marquis, in an extraordinary fit of generosity, gave the specimen to his friend, the eminent surgeon and keen naturalist, Sir Philip Crampton. In the same year the British Association held its annual meeting in Dublin, hosted by the Zoological Society of Ireland. Sir Philip, who was on the committee of the Society (he later became its President), brought the fossil with him to Dublin, to exhibit at the meeting. The Society's report shows that they did not have a building big enough to hold it.

> "It is stated that in the coming summer a collection of reptiles may be formed, members could assist in this by asking their friends in foreign countries to send home

*The largest fossilised plesiosaur found on the Yorkshire coast, as depicted in 1863.*

specimens, which, for the most part, will come safely if put in a box having a few air-holes, and placed in the hold of a ship where they may not be subject to much change of temperature. The reptiles may be appropriately managed in a house which is now in course of construction for the finest skeleton of Plesiosaurus known. It was presented by the Marquess of Normanby to Sir Philip Crampton, who has kindly placed it at the disposal of the Society for exhibition. This most interesting relic of one of 'the great Sea Dragons' of the ancient world, will, no doubt, prove eminently attractive, not only to the citizens of Dublin, but to the many scientific and other visitors likely to visit this city during the next few months. Its size is so great that the Council felt obliged to construct a special building of 36 feet long for its due exhibition. The building will soon be completed, and the skeleton and illustrative specimens prepared for inspection in it."

Annual Report, Zoological Society of Ireland, 1853

Six years later Sir Philip Crampton had died, and ownership of the plesiosaur passed to his son.

"With the kind attention of Messrs. Jukes and Sanders, the Council have taken measures to protect, and they hope to permanently preserve, the skeleton of the Plesiosaurus, a specimen of the fossil world unrivalled in size and perfection. This wonderful specimen of a long past age, the Society is aware, was placed in the

Gardens by your late lamented President, Sir Philip Crampton, and at his death was presented as a donation by his son – the present Sir John Crampton."

Annual Report, Zoological Society of Ireland, 1859

The fossil was first given a full scientific description in 1863. The paper by Carte and Baily in the *Journal of the Royal Dublin Society* gives a history of the specimen, followed by a detailed description of the skeleton. The palaeontologists then name the specimen after its former owner.

"After as careful an examination as possible, and comparison with other specimens, we consider this saurian to be a new species, and have therefore named it *Plesiosaurus cramptoni*, as a slight tribute to the memory of that eminently scientific man." *

In the second half of the last century the fossil changed ownership within Dublin's scientific institutions. It was loaned to the Royal Dublin Society in 1863 for display in their museum. The museum was incorporated into the state-run Museum of Science and Art (later called the National Museum of Ireland, now known as the Natural History Museum) in 1877, whereupon a nominal sum of £200 was paid to the Royal Zoological Society of Ireland to 'secure clear ownership'. So far so good, but the history of this magnificent fossil (specimen number NMING F8785) in the twentieth century has been less fortunate. In a letter to the

* In 1874 H G Seeley introduced the name *Rhomaleosaurus* for those pliosaurs with a particular pelvic structure. This species was taken as the type species of the genus and renamed *Rhomaleosaurus cramptoni*. I am grateful to Nigel Monaghan for this information.

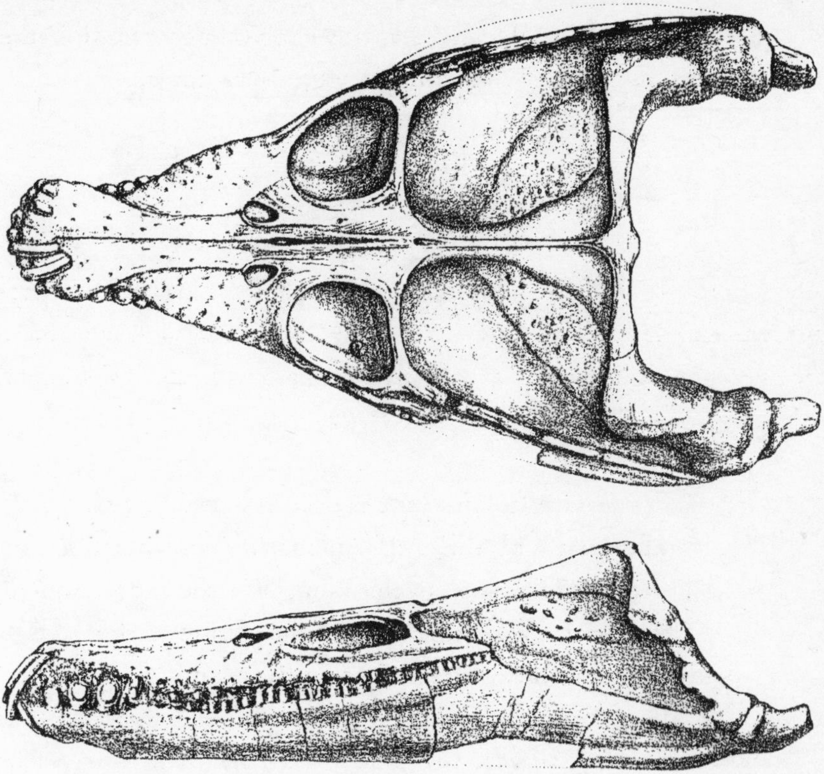

*The head of* Rhomaleosaurus cramptoni *seen from above and the side.*

author, Nigel Monaghan, Curator of Geology at the National Museum of Ireland, outlines the subsequent history of the specimen:

> "In 1890 the museum's fossil collections were transferred to a neighbouring building referred to as the fossil hall. Here it lay until 1961, but since 1922 the fossil hall had been closed to the public for security

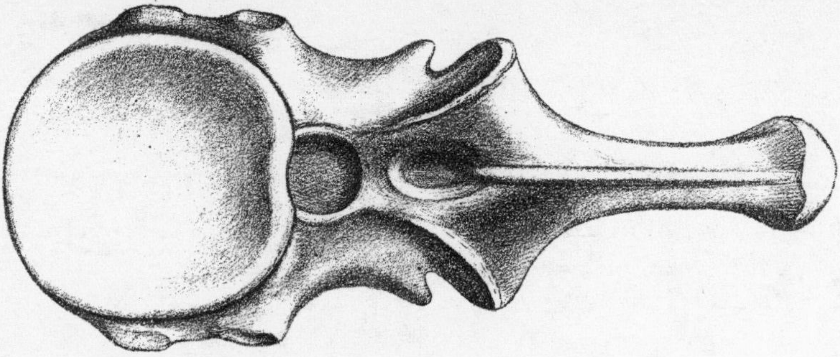

*Plesiosaur vertebra from* Rhomaleosaurus cramptoni.

reasons as it was beside the new Free State parliament building (now Dail Eireann) in Leinster House.

At various times from about 1930 to 1960 the fossil hall was open to the public, but in 1961 the Government decided that the space occupied by the building was needed for a restaurant and office block. The fossil hall was demolished and the collections moved to storage in the Royal Hospital in Kilmainham, on the west of the city (now the Irish Museum of Modern Art). The geological collections were transferred to Daingean, Co. Offaly in 1979 and were moved again in 1992 to our reserve stores ... a mile from our offices at Beggars Bush, Dublin 4."

For those who want to see the specimen, there is bad news and good news. The bad news is that the body of the fossil is broken up and kept in crates in storage, though the skull is in the National Museum of Ireland's laboratory, where it has undergone conservation work. The good news is that during the last 140 years several casts of the specimen have been taken, and these can now be seen in the Natural History Museum in London, Bath Literary and Scientific Institution and Cornell University in New York. The Natural History Museum remounted its display of fossil reptiles, after extensive cleaning and conservation work, in 1993. The cast of *Rhomaleosaurus cramptoni*, the grandest fossil reptile specimen ever found on the Yorkshire coast, forms a magnificent centre-piece to this display.

chapter eighteen

# Embracing and Uniting

## *Geology and classification*

A system of classification of plants and animals (both fossil and living) that can be universally applied is of central importance to natural science. Even if we draw back from the view that natural science is nothing more than the relentless working through of the task of classification, we can agree that science cannot proceed beyond local or regional curiosity without general systems. In the early nineteenth century, as the speed of new discoveries threatened to overtake the adoption of universal concepts, the problem of classification became acute. Scientists, being pragmatic, realised that a solution was urgently required. One system of classification began to be favoured above others, and rapidly became the most widely adopted. Martin Simpson, Curator at Whitby Museum for most of the second half of the nineteenth century, and one of the pioneers of fossil taxonomy, explains the difficulty:

> "In drawing up a Descriptive Catalogue of the Fossils of the Yorkshire Lias, I without hesitation adopt the views of the late Baron Cuvier, which I consider to be by far the most comprehensive, the simplest, and the most instructive, of any which have yet been promulgated on the classification of animals. Whilst the system of Cuvier, founded upon the strictest, and most perfect anatomical investigations, embraces and unites, both the systems of Aristotle and Linnaeus, his principles of nomenclature, seem to me the best calculated to rescue natural history, from that confusion

into which it has unhappily fallen. By adopting the comprehensive generic groups of Linnaeus, as genera, under which species are to be united and named, thousands of names which now becloud and perplex natural history, might be rendered either very subordinate, or entirely swept away. And by adopting such genera as those of Lamarck, as sub-genera, or divisions of genera, the species can be the more easily described and understood. Also by this method naturalists having different views of nomenclature, might still understand each other. For whilst the followers of Cuvier and Linnaeus would adopt the names of the more comprehensive groups, as the names of their genera, others might prefer the names of smaller groups, for generic names, as those of Lamarck; yet both parties would understand each other; the sub-genus being merely a division of the genus. If, however, naturalists will go on multiplying generic names of but little importance, in the great system of nature, and often worse than synonymous, the prediction of Cuvier will be verified: – That the advantages of the binomial system so happily imagined by Linnaeus will be lost; and geologists, and men of general learning, will in a great measure be excluded from pursuing the details of Natural History; the number of genera being already far beyond any ordinary man's comprehension."

chapter nineteen

# The Eighth Reptile

## *A question of length*

In the 1860s the cliffs stretching from Loftus to Blea Wyke were being quarried for alum, in some places for iron ore, and in some places for a curious opaque black stone called jet. In the early part of that decade (the exact date was not recorded), on the cliffs below the village of Hawsker, to the south-east of Whitby, some jet workmen came upon a large fossil. It turned out to be a specimen of *Ichthyosaurus crassimanus*. The fossil was bought by the Whitby Literary and Philosophical Society, for display in their museum (WM25468), for £105. With a length of 25 feet and a width of eight feet across the extended paddles, it was the largest complete specimen of *I. crassimanus* then known.

Around seventy years later, in the 1930s, the Society raised enough money for the building of a new museum at Whitby, to replace the over-crowded premises they had occupied for the past hundred years. The museum was designed to accommodate the Society's treasures and to allow room for more acquisitions. The following account was found among the papers left by the late curator of the museum, Frank Sutcliffe, who is known to the world for his photographs of the working people of Whitby:

> "It is odd how the Goddess of Chance often arranges things better than the most skilful designer. When it was decided to build a new museum at Whitby for the better housing of its varied treasures, the question of displaying to the best advantage the seven fossil Saurians, scattered here and there in the old museum,

was one of the main problems; for, as everyone knows, the main feature of Whitby Museum is its collection of Lias Fossils, gathered and arranged with such great skill by Martin Simpson.

The largest of the Saurians, *Ichthyosaurus crassimanus*, was measured by the architects of the new building which was to be added to the Art Gallery in Pannett Park, and the walls were made large enough to take it; in fact the size of the new building was determined by the length of this fossil reptile. Then, when the beast had been cut up into four parts, he was too big to be removed whole through the windows of the old place; and, when ready for fixing up on one of the new walls, it was found that he was longer than the wall. Either he had stretched a foot or so; or the wall had shrunk on drying; or – tell it not in Gath – the tape measure used in taking his measurement had lost some inches, or had shrunk in washing.

I forgot to say that before any of the Saurians were removed from the old walls where they had hung for nearly a hundred years, paper patterns of their full size were made of them all. So, when the paper pattern of *crassimanus* was stretched across the wall where it was decided he should be placed, it was found to be too long. Then arose the question: should a foot or so be cut off the end of his tail, or should he be placed diagonally instead of horizontally. The counsels of the wise philosophers decided on the latter course; and now everyone who sees the beast says how well he looks, and how much more alive he appears than if he had been on a dead level."

chapter twenty

# Thinking in Four Dimensions

## *The problem of the Whitby fault*

Stratigraphy is the most sublime of sciences, as well as being the most mind-boggling. The re-imagination and re-creation of the history of the Earth's crust, through the study of its present intricate formations; a four-dimensional jigsaw with most of the pieces missing – now there's a game that's worth the candle. But also one with its own dangerous seductions. The fourth dimension, time, can easily become, through its convenient unknowableness, the repository of solutions to serious problems, and the first and obvious refuge for confused minds. But the obvious is hardly ever the truth, and never the whole truth, and time is not always the best place to look for history.

When you next go to Whitby pay your £2 on the west pier, and take a twenty-minute trip in the old Whitby lifeboat, the *Mary Ann Hepworth*. From the tranquil water of the outer harbour, she will take you out through the jaws of the curved breakwaters, heading for the buoy that marks the seaward end of the deep-water channel. As you hit the open sea the boat will pitch and roll – keep your eyes on the horizon to avoid seasickness. Boats should keep to the west of the buoy to avoid the rock shelf that runs out from the shore. But on a high tide they will take you round the east of the buoy, before turning south-west toward Sandsend, and then tracking back east, parallel to the beach, to complete the triangle. On the outward leg, turn and look back at the town. Avoid becoming entranced by the picturesque confusion of the roofs and walls climbing out of the narrow harbour; draw your eyes away from the abbey ruins;

*Whitby outer harbour from the air. The Whitby fault line is thought to follow the deep water channel of the River Esk.*

look instead at the cliffs on either side of the harbour. To the east a towering sombre mass of finely bedded grey shales, with a thin layer of lighter sandstone at the very top. On the west side a low series of yellow sandstone cliffs, made out of huge square blocks. The East and West Cliffs are as dissimilar as chalk and cheese. The attempts to explain this difference are an illustration of the niceness of stratigraphical thinking, and how it can go wrong. From the birth of the science in the second decade of the nineteenth century, until the 1940s, geologists saw the cliffs as a two-dimensional puzzle, in which time played the role of master builder. Each worker seemed to 'improve' on his predecessor's work. But they were progressively getting things more and more wrong. It took 130 years for the mistake to come to light, and then the real explanation for the peculiarities of Whitby's cliffs had to be figured out.

The Reverend George Young, pastor of Cliff Lane Chapel in Whitby, was a pioneering figure in the golden age of amateur cleric naturalists. He wrote a biography of Whitby's most famous son, Captain Cook, a history of his adopted home town, the first book on the geology of the Yorkshire coast and a two-volume book on scriptural geology, among other works. He was also the driving force behind the founding of the Whitby Literary and Philosophical Society and its museum. He was the first to record the presence of a fault at Whitby harbour:

> "Besides the numerous veins and vertical fissures that cross the strata in our hills, and the frequent undulations of the strata already noticed, some remarkable interruptions occur which demand observation. At the mouth of the Esk, a *slip* or *downcast* has taken place on the north side,* the whole mass of strata on that side being 80 or 100 feet lower than the corresponding strata on the south side; and this interruption seems to be continued through the whole vale of the Esk."
>
> George Young, 1817

In his geological survey Young goes into more detail, and notices a peculiarity of the fault – it disappears once you move inland:

> "At the mouth of Whitby harbour, in the bed of the Esk, we come to the second great slip in these strata.

*Though Whitby lies on England's east coast, the town actually faces north. The cliffs are more usually and correctly known as the West Cliff and the East Cliff, though Young and other writers sometimes refer to them as the North and South Cliffs respectively.

The alum shale proceeds to the middle of the current, more than half way across the harbour, and there terminates abruptly, nothing but sandstone rocks being found on the other side. To what depth the strata on the west or north-west side of the harbour have sunk, cannot be very correctly estimated, especially as the sandstone beds are so variable, that it is difficult to trace the correspondence between those on the one side of the slip, and those on the other: but as the sandstone beds that come down to the beach at the battery appear to correspond, whether with the highest beds of the east cliff, or with a higher portion of the sandstone there wanting, we can scarcely reckon the amount of the slip less than 100 feet.

In attempting to trace the progress of the break to the south and west, we find ourselves lost amid the irregularities of the strata. It appears sometimes to follow the course of the Esk and sometimes not . . . Immediately beyond the harbour, the strata rise again on both banks, and as the sandstone cliffs correspond, we see no vestige of the slip."

George Young and William Bird, 1822

John Phillips spent his youth travelling the north of England with his uncle, William Smith. Although Smith was the man whose ideas and discoveries founded the science of stratigraphy, he was by instinct a practical man, not a writer or a teacher. So it was his nephew who became the Curator of the Yorkshire Museum and then Professor of Geology at Oxford. And though Smith often threatened to write a book on the geology of the county where they both lived, it was Phillips who brought the idea to fruition. Though Phillips found George Young's earlier account to be inaccurate in some places, he agreed on the importance of the Whitby fault:

> "Proceeding from High Whitby, the cliffs fall gradually toward the north, and at the same time the lias rises to the height of one hundred feet above the sea. It however sinks again near the harbour at Whitby, where a great dislocation depresses it suddenly on the north, and prevents its distinct reappearance as far as Sandsend . . . The great dislocation . . . extends a considerable distance up the valley of the Esk. Its effects are very remarkable at the sea side. On the south side of Whitby harbour, a part of the cliff is composed of the upper alum shale, and this rock extends far into the sea, making broad level scars at low-water; but on the north side of the water is a cliff of sandstone and a beach of sand. The exact amount of the depression occasioned by this fault cannot, perhaps, be determined; but I estimate it to be not less than one hundred and fifty feet."
>
> John Phillips, 1829

Once the basic geology of most parts of Britain had been described, there came a host of books aimed at walkers, tourists and amateur naturalists and visiting geologists. These were not scientific investigations, so they simply repeated the conclusions drawn by their predecessors:

> "At Whitby, a fault as already mentioned running up the lower part of Eskdale, has thrown down the Lias

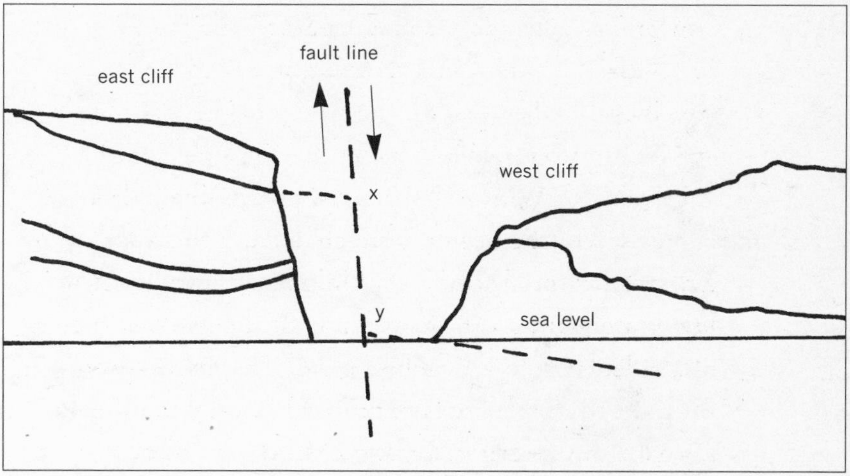


*The section from John Phillips's book of 1829 (top) shows how the sandstone bed on the top of the East Cliff (whose thickness is exaggerated) matches the sandstone of the West Cliff. The bottom diagram shows how this works in practice. x marks the extrapolation of the base of the East Cliff sandstone, and y is the highest possible point of the base of the West Cliff sandstone, i.e. sea level. The vertical distance from x to y is about 150 feet. This is the minimum movement along the fault line. If the base of the West Cliff sandstone lay below sea level, then the drop across the fault would be correspondingly greater. The sections are as viewed from the sea.*

beneath the sea level. This causes the very remarkable difference between the cliffs on opposite sides of the harbour. On the east side is a lofty cliff, 185 feet high, of Lias capped with the sandstones of the Lower Oolite, at the top of which, looking over a wide stretch of sea on the one hand, and of moorland on the other, stands the conspicuous and ancient pile of Whitby Abbey. On the west of the harbour is a comparatively low, waterworn, sandstone cliff belonging to the Lower Oolite, covered with glacial drift. On this side the beds have been thrown down by at least 150 feet."

Charles Bird, 1881

A popular tourist guide from the turn of the century mentions an intriguing aspect of the Whitby fault. Because it runs beneath the harbour, no one has ever seen it. It exists only as an inference:

"In the bed of the River Esk occurs another remarkable fault. It is quite beyond the reach of examination, though its consequences are evident enough. There must be a downthrow of at least one hundred and twenty feet, for it will be seen that the Sandstones which form the summit of the East Cliff, are exposed only some few feet on the West Cliff, and quickly descend below the shore westward of the Saloon, and do not reappear till close to Sandsend."

*Horne's Guide To Whitby*, 1904

In 1835 the Geological Survey was founded with the aim of describing and mapping the geology of the whole of Britain to a professional standard. Charles Fox-Strangways joined the Survey in 1867 and began work on the Yorkshire Dales. He is best known for his work on the Jurassic rocks of the Cleveland area and the chalk wolds further south. His monument is *The Jurassic Rocks of Britain,* published in two volumes by the Geological Survey in 1892. His detailed work on the country between Scarborough and Whitby was published after his death in 1915. Fox-Strangways recognised that a narrow and distinct layer of rock, known as 'the Dogger' bed, held the key to understanding the Whitby fault, but he followed the error of his predecessors – in fact he increased the magnitude of their mistake:

> "At Whitby there occurs an important fault, known as the *Whitby Fault,* parallel to that of the Peak and presumably of the same age. Its displacement may be judged by comparing the two sides of the gorge-like mouth of the river. On the east side the Dogger is at about high-water mark. On the west side the Grey Limestone is only a short distance back from the top of the cliff. As the base of this bed is about 450 feet above the Dogger, and the cliff is some 200 feet high, the downthrow to the west is rather more than 200 feet. The seaward position of the fault is shown by the truncation of the Jet Rock, which starts from the foot of Saltwick Nab and runs westward as a sunken reef, marked by a line of breakers that ceases abruptly at the line of fault."

C. Fox-Strangways and G. Barrow, 1915

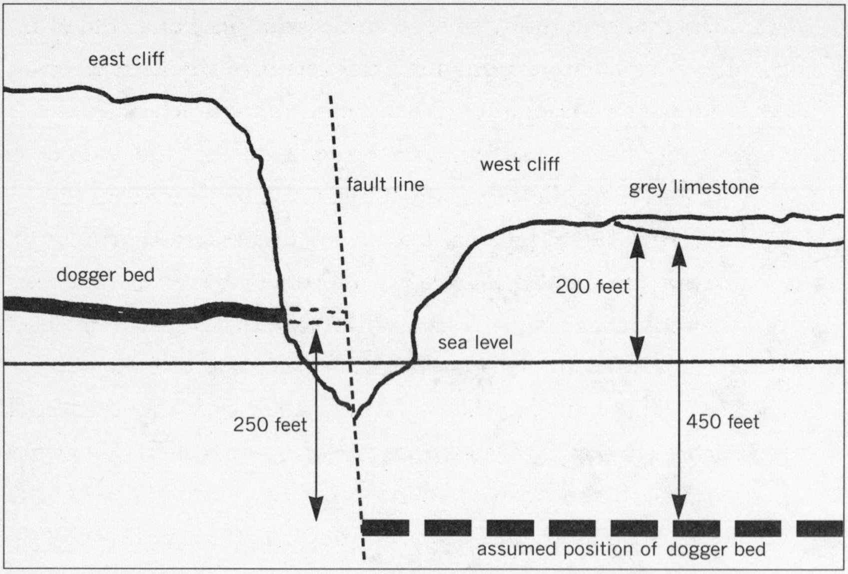


*This section shows the thinking behind Fox-Strangways's estimate of a 250-feet downthrow to the west across the Whitby fault. His experience in mapping the area led him to believe that the key Dogger bed always occurred 450 feet below the grey limestone. Once he found grey limestone on top of the West Cliff sandstones, a simple mathematical deduction showed the difference in heights of the Dogger bed on each side of the fault.*

Leo Walmsley was a local writer whose novels about the working people of the Yorkshire coast brought him national prominence in the early decades of the twentieth century. His books were greatly admired by T. S. Eliot. Walmsley was a keen amateur geologist. He introduces the notion that without the Whitby fault, there would be no Whitby:

> "The Estuarine Beds form the upper part of the cliff under the Whitby Abbey [the east side]; while at the other side of the river [west side] they may be seen composing the cliffs in Khyber Pass and under the Spa.

> At this side they are seen to be crumpled and folded in an extraordinary manner, the result of the thrusting and shearing which took place when the whole series of rocks was fractured, and the beds on the west side of the river sank down something like one-hundred-and-twenty feet. Had it not been for this fracture or *fault*, it is probable that Whitby would never have existed, for, even as Whitby's origin and prosperity has depended on its river and harbour, so has the existence of its river and harbour depended upon the line of weakness which this fault presented to the eroding actions of the ancient glaciers and post-glacial floods."
>
> Leo Walmsley, 1919

Percy Fry Kendall retired as Professor of Geology at Leeds University in 1922, and immediately wrote and published his two-volume *Geology of Yorkshire*. Kendall, an expert on glaciation, knew that Whitby harbour was a relatively recent ice age creation:

> "The present mouth of the Esk is not that by which the river flowed to the sea in Pre-glacial times. The precipitous bluffs of rock forming the Khyber Pass on the west side of Whitby and the quaint and picturesque 'yards' of Church Street and Henrietta Street on the east side proclaim to the eye of the geologist that the gorge of which they form part is no very ancient valley. The old estuary of the Esk is indeed to the westward, and it is indicated on the shore about Upgang by a long stretch where all live-rock disappears and the cliffs and beach are wholly composed of boulder clay.

> The original topography was entirely buried and obliterated by the invading ice, and when the waters once more found an outflow in this direction they settled upon a line of weakness in the solid strata and carved out the ravine which serves Whitby as a harbour."

But Kendall did not then wonder if the Whitby fault was as great as had been supposed. He recounted Fox-Strangways's view, which carried the imprimatur of the Geological Survey:

> "Whitby Harbour was formed in Glacial times along the line of a disturbance. Stand at the foot of the cutting called the Khyber Pass, where estuarine sandstones – strongly current bedded – are seen lying in a trough of the shales, – and after examining these look directly across the harbour. Below the old church is Upper Lias surmounted by the Dogger and Estuarine sandstones. Obviously the west side of the harbour is relatively downthrown. The displacement is estimated by the Geological Survey as 200 feet, and is attributed to a fault."
>
> Percy Fry Kendall and Herbert E. Wroot, 1924

We now encounter the twentieth century's expert on the geology of the Yorkshire coast, and the man who was to unravel the problem of Whitby harbour, J. E. Hemingway. Hemingway was a native of Whitby who became Professor of Geology at Leeds and President of the Yorkshire Geological Society. He worked on the key Dogger bed at Whitby but initially followed the earlier conclusions about the Whitby fault:

> "At the Spa Ladder on Whiby East Pier the Dogger, here 40 feet OD, is 2ft 6in thick, and is deeply weathered and overlain by shales with rows of small ironstone nodules. From this point the outcrop swings south into Whitby Harbour, dipping steadily to river level.
>
> The Whitby Fault, with a downthrow of 200 feet to the west, throws the Dogger below sea level for two miles. It reappears by the roadside near East Row, Sandsend, where the usual sideritic sandstone is overlain by a black oolite."
>
> R. H. Rastall and J. E. Hemingway, 1940

There is then a crucial nine-year gap when no significant new work on the geology of Whitby was published. During this time Hemingway came to realise that all the previous explanations of the differences in the East and West Cliffs were mistaken. Hemingway identified three elements – structural evidence, palaeobotanical evidence, and the presence of a reef of the Dogger bed on the west side of the harbour, which made him change his mind. Re-reading the above extract from his 1940 paper we can see that the third element, the Dogger reef, was a crucial anomaly. If, 'the outcrop [of the Dogger] swings south into Whitby harbour [on the east side], dipping steadily to river level', and there is then a fault with a downthrow of 200 feet, there could not be any Dogger on the west side. It should, as Hemingway earlier thought, be buried underground until it emerges two miles west at Sandsend. Putting things the other way round, when Hemingway found a Dogger reef below the high tide level on the west side of the harbour, he knew that the fault could not be very large. But then he realised that if the throw of the fault was less than 120 feet, it could not account for the difference in rock types between the East and West Cliffs – there had to be another explanation. The first published indication of Hemingway's volte face, and his surprising conclusions, is a report of a field trip:

> "During the evening preceding the excursion, 26 members met at the Clarence Hotel, Whitby, to hear a short account of the geology of the Whitby area by the Director [Hemingway]. With the aid of structure contour maps, drawn on the Dogger, the basin structure of the area was demonstrated. The position and throw of the Whitby fault was also shown with greater accuracy than had been previously possible . . . On the Saturday, the party walked to the East Pier for a general view of the harbour and cliffs. It was emphasised that despite the dissimilarity of facies in the Lower Deltaic Series to the west of the harbour (mainly sandstones) and the east (mainly clays), the structural and palaeobotanical evidence demonstrated their uniformity of age . . .
>
> Near low water mark, a 200 yard reef of Dogger sediments was exposed, showing typical Black Oolite which also occurs near Sandsend. This exposure is critical to the structural interpretation of the district."
>
> J. E. Hemingway, 1953

Hemingway had decided that the East and West Cliffs were exactly the same age, and had formed at the same time. The palaeobotanical evidence came from Hemingway's colleague at Leeds, Tom Harris. He refers, in a 1953 paper, to earlier work he had done on the fossil plants of the Whitby strata. Unfortunately he does not tell us which plant species provided the key to the Whitby problem, and the Whitby fault is not referred to in his

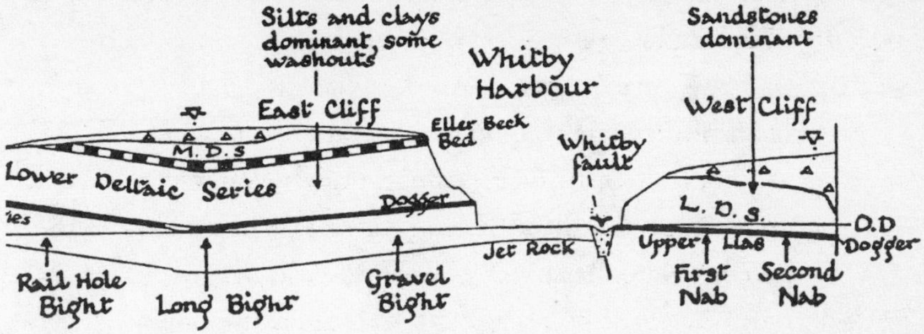


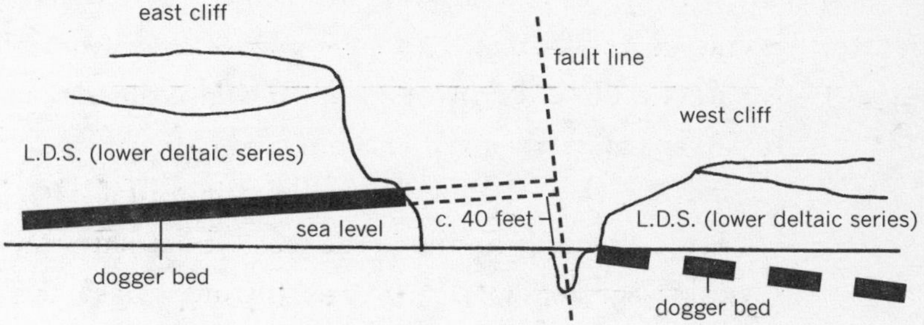


*Once Hemingway discovered a reef of Dogger just below sea level on the west side, he knew that a new interpretation of the Whitby fault was needed. The shales on the east side and the sandstones on the west are both labelled 'L.D.S.' or Lower Deltaic Series on his section, showing that they are of the same age. The second section is a simplification showing the drop of the Dogger bed across the fault.*

earlier work. Since Hemingway's detailed mapping of the Dogger also remains unpublished, we can only assume that conversations between Hemingway and Harris, some time between 1942 and 1949, persuaded them that the Whitby cliffs were of the same age.

> "There is a small group of species which occurs in the Lower Deltaic but nowhere else; and though none of

> these is very common they have from 5–20 localities to their credit so that it is possible to feel some confidence that they are real indicators of this stage. I was able to use these species as evidence that the rocks at Whitby West Cliff are Lower Deltaic and that there is no very considerable fault in Whitby Harbour. More often than not, however, these characteristic species are lacking."
>
> T. M. Harris, 1953

Once he was sure of the evidence, Hemingway could reflect on his predecessors' misinterpretations. But what of the Whitby fault? Did it still exist and, if so, how significant was it?

> "The Whitby fault has been regarded as an important structure running from the Battery along the harbour to Spital Bridge and dying out south of Stainsacre. Its throw was estimated at 200 feet. The significance of the fault has, however, been exaggerated. Its throw at Scotch Head is not more than 40 feet. It dies out rapidly to the south and does not continue south of Spital Bridge. It probably represented a line of weakness in the rocks followed by the post-glacial river and may in consequence be a factor in defining the position of Whitby Harbour. Apart from this it is of no importance."
>
> J. E. Hemingway, 1958

Hemingway was asked to write the Geologists Association guide to the Yorkshire coast – a book for the amateur geologist – in 1963. He refers again to (his own?) detailed mapping of the Dogger, and to borehole evidence, but unfortunately does not give any references for these:

> "The east Cliff exposes forty feet of Alum Shales (Upper Lias) at its base, succeeded by a thin representative of the Dogger (2 1/2 feet), which in its turn is followed by the Lower Deltaic Series (103 feet). This latter is dominated by fine sandstones, silts with rootlet beds and carbonaceous clays in cyclic succession. It is followed by the marine Eller Beck Bed (c. twenty feet) which is capped by boulder clay . . . By contrast with the East Cliff, however, the West Cliff and Khyber Pass are made up of massive yellow and yellow-brown sandstones, with only subsidiary silts and clays. Because of this striking difference, as well as the westward termination of the submarine Whitby Rock along a line nearly due north of Whitby Piers, and the position of the River Esk itself, a fault with a westerly downthrow of up to 200 feet was postulated along the deep channel of the harbour. The West Cliff succession was thus assigned to the Middle Deltaic (Estuarine) Series. Detailed mapping (particularly of the Dogger), as well as borehole evidence within the harbour shows that the fault, which is not exposed, downthrows not more than forty feet to the west and is relatively unimportant in the development of the River Esk.
> The West Cliff succession is a part of the Lower Deltaic Series, a conclusion which has been confirmed by spore-analysis."

J. E. Hemingway and V. Wilson, 1963

In a more technical publication, Hemingway explains how two different rock types can be formed alongside each other at the same time. The area was a low-lying coastal marsh with the outflow, or distributary, channels of a large delta passing through it. Dark shales were forming at the bottom of the stagnant saltwater marshes, while pale sandstones formed in the channels of fresh oxygenated water being fed by river water, that carved their way through the marshy, half-solidified mud. Occasionally the levee, or raised bank, of one of these freshwater channels would collapse, dumping a pile of accumulated sand into the stagnant mud. 'Facies' is a catch-all term for geologists, but essentially means a rock formation whose constituents formed at the same time under the same conditions. A 'facies change' is therefore a change from one rock type to another, reflecting a change in origins. The 'clastic range' is the spectrum of different rocks that are formed from eroded pieces of other rocks, and includes sandstones and shales. Do not be too hard on geologists for using these exclusive terms – they needed them to cope with the changes in their subject. The formation of sedimentary rocks was much more complex, and interesting, than they had previously understood.

> "Facies change within the clastic range is frequent, but can only rarely be related to the deposits of a distributary channel. Thus rapid thickness and facies change may be the product of levee breaks, as in the Whitby cliffs both to the east and west of the town. Indeed these sections, within a distance of a kilometre, are an epitome of delta-marsh and washout deposition as expressed in the Yorkshire Jurassic Basin. Nevertheless, evidence of lateral passage from a sand-filled distributary channel to the adjacent marsh sediments is rare. The more usual sharp contact demonstrates erosion of the channel into marsh

> sediments that were compacted but not lithified, rather than simultaneous deposition within the typical distribution channel with interdistributary marsh infilling."
>
> J. E. Hemingway, 1974

A contemporary textbook followed Hemingway's estimate of the size of the Whitby fault. The fault was shrinking, and would shrink still further:

> "Viewed from one of the cliff tops Whitby, with its distinctive red tiled roofs crammed tightly on both sides of the River Esk, is one of the most picturesque coastal towns in England. The harbour channel is a shallow gorge-like feature and appears to follow closely the line of a fault which has a throw of about 40ft (12m)."
>
> Derek Brumhead, 1979

Some geologists realised that work done in physics and engineering on patterns of flow could be used in geological problems. The physical environment that existed at the time the rocks were formed could be recreated as a 'model'. The direction of the flow of rivers, tides and currents across the marsh, or beach, or coastal sea-bed could be worked out from the composition and physical features of the rocks, for example ripple marks, and the way particles are graded by size within the rock. The 'model' is then mathematically run to see how it fits with the geological outcomes. 'Pedogenic features' are things in mudstones that tell you where the soils they contain come from. Jan Alexander estimates that the difference in height at the Earth's surface which was caused by the fault (the topographic expression) was small:

> "The most dramatic facies change is seen at Whitby where West Cliff is dominated by stacked channel sandstone bodies (comprising over 80% by volume) whereas East Cliff, the footwall has sandstone percentages around 20%. The sediments of East Cliff consist of sheet sandstones of flood origin interbedded with siltstones and mudstones with abundant pedogenic features.
>
> There is, at present, no way of correlating individual facies across the observed structures but the amplitude of incision and the slight changes in pedogenic conditions suggest that the topographic expression was never more than a metre or two."
>
> Jan Alexander, 1986

Finally we enter the world of soil chemistry, for a confirmation of what was going on when the entirely different Whitby West and East Cliffs were forming right next to each other. 'Catena' is a nice term for geochemists. It means a whole group of soils that were formed from the same parent material; but each may be different in composition. This difference can then be explained by, and is the key to discovering, variations in drainage patterns or surface levels in different places. The Whitby cliffs show that small changes have dramatic consequences. A final flush of technical terminology – a 'ganister' or 'seatearth' is a sediment that contains plant rootlets; a 'crevasse splay sandstone' is one that forms after the collapse of a river bank or levee; 'fluvial aggradation' is the deposition of silt and sand by rivers:

> "In the non-marine sediments lateral (catena) variations in soil profiles can be related to floodplain aggradation

and abandonment. Near the major channel sands in the Saltwick formation at Whitby West Cliff, a ganister and seatearth formed in elevated and well drained sediments that were frequently flushed with fresh water. . .

These strata have been buried and then uplifted twice since [their formation in] the Middle Jurassic. They were buried rapidly to almost 1500m during the late Jurassic to early Cretaceous, and then uplifted to within 100m of the surface during the mid-Cretaceous. The sequence was reburied to more than 2000m during the Tertiary before renewed uplift and erosion, resulting in exposure at the outcrop today . . .

The mineralogy of the crevasse splay sandstone and associated shales at Whitby West Cliff reflect prolonged fresh water flushing during a temporary depositional hiatus and plant colonisation . . . The overbank sands and shales in the Saltwick Formation at Whitby West Cliff contain a well-developed ganister and seatearth. These formed as oxygenated fresh water from nearby channels percolated through the sands and into the shales. Suboxic ferric iron reduction led to the formation of siderite ironstones at . . . other areas of the Saltwick formation [e.g. the East Cliff] with fewer channels. This catena relationship arose because fluvial aggradation raised areas of the floodplain and preferentially flushed them with oxygenated fresh water."

J. D. Kantorowicz, 1990

So, the mismatch of the Whitby cliffs came about because entirely different sediments were deposited alongside each other at the same time. Freshwater channels (giving yellow sandstones) carved their way through stagnant half-solidified mud banks (giving dark shales). The same thing is happening right now at the mouths of the Nile, the Mekong and the Mississippi.

The understanding of the Whitby fault went wrong because it was such a triumphantly easy problem to solve (wrongly) in the early days of geology. It then ceased to be a problem and was of no intellectual interest for 130 years. We think we know better now, and we probably do. But every so often someone should have a look at Harris's plant specimens, and at Hemingway's Dogger maps, just to make sure they really do tell us what we assume they tell us.

chapter twenty-one

# The Ninth Reptile

## *An American in Whitby*

The exact date of discovery of the type specimen of *Rhomaleosaurus longirostris,* one of the largest plesiosaur species, is not known.* In 1876 it was in the possession of Mr Brown Marshall, a carpenter by trade, but better known as one of Whitby's great fossil-finding family. The specimen came to the attention of Professor John Potter Marshall (no relation) of Tufts College, Massachusetts. Professor Marshall was on a study tour of Europe, which then turned into a buying trip. He recorded the events surrounding his purchase of the plesiosaur in a letter to the board of his college, in which an American professor proves a match for a Yorkshire fossil dealer (at least by his own account):

> "Prof John Potter Marshall, College Hill, April 29, 1876
>
> To the Honorable Board of Trustees of Tufts College
>
> Gentlemen
> When I arrived in London on my way home, last August, I learned for the first time that money had been appropriated by you, for the purpose of purchasing fossils and apparatus, for the college. My plans for improving the remainder of my time before sailing on

* A type specimen is one which is agreed to embody all the essential characteristics of a species, and which is therefore used as a yardstick for the classification of other specimens.

the 18th of that month, were to spend a few days in visiting the museums of Nat. History in London, and then with a friend, who was waiting at Liverpool, to make an excursion among the mountains and mines of Wales. These plans had to be given up.

As I could not draw upon McCalmont Brs. & Co. until the expiration of a week from the time the drafts had been deposited with them, and as I could not purchase fossils at Whitby, without money in hand, I was obliged to wait in London a week, which time I employed in selecting apparatus. A considerable portion of the apparatus was to come from German & French makers, but I decided that even if I had time to go back to Darmstadt and Paris, (which I had not) the expense would be no more if ordered through Elliott Brs. of London.

I selected carefully the apparatus, ordered of the London makers, and at the close of the week made a journey to Whitby in the north of England, having previously telegraphed and obtained a refusal of the Plesiosaurus.

Late on Tuesday night I arrived at Whitby. On Wednesday, early in the morning I called on B. Marshall, the owner of the fossils and began negotiations.

He was very firm, as to the price of the Plesiosaurus, £150, *cash* in *hand*, and showed me a letter from Alpheus Hyatt, Custodian of the Museum of the Nat. History Society of Boston in which the offer of £140 was made for the fossil delivered in Liverpool.

After a large part of the forenoon had been spent in

bargaining, he agreed for £150 to deliver the Plesiosaurus in Liverpool free of expense to us.

I then proposed that a carpenter should be engaged to make the boxes, of which seven would be required for packing all the fossils I proposed to purchase. He said nothing could be done that day and the next as there was a great Cattle Fair in Whitby and the carpenters would not work. He told me that I could safely pay him for the fossils and he would send them to Liverpool all right. I have no doubt he would have done so, as he is a very respectable and well known dealer in jet goods and fossils – but I did not like paying out so large a sum before the delivery of the articles. So after a time, by the help of his wife, I prevailed upon my gouty acquaintance to look up some carpenters. It was late in the afternoon before the boxes were begun, but by some personal exertion, I succeeded in getting the Plesiosaurus and smaller fossils all packed and loaded upon the railway dray late in the evening of Thursday.

Had there been ample time to get the large Ichthyosaurus off, I should have been tempted to have purchased it. Still on the whole I am confident I did the wiser thing in purchasing a large number of smaller fossils which can be used to great advantage in giving instruction in the college. As you see by the list, I brought home a small but excellent Ichthyosaurus, 8 feet long, which will answer our purpose for the present.

The Plesiosaurus is a fine acquisition being the most perfect of its size that has ever been obtained. I saw on the beach at Whitby the spot, covered at high tide, where the bones were found.

Mr. Marshall furnished the Plesiosauri to the British and York Museums & his father contributed the one in the Whitby museum. I have seen them all and know that ours is a far finer specimen, than any of these famous fossils.

By the bill of Ross, you will see that I purchased a microscope of him. He is acknowledged to be one of the best if not the best maker in the world. He assured me that for certain reasons, he desired to have one of his first class instruments near Boston, and would spare no pains to give the College as perfect a microscope as he could make.

I am happy to assure you that it proves to be all that can be desired and I feel sure that you will approve of my action, in devoting a portion of the £300, you authorized me to spend for my department, to the purchase of an instrument worthy of our college.
The one we had before, made by Hartnecht of Paris is very good so far as it goes, but such improvements have been made within the past fifteen years, that it can no longer rank as a first class instrument.

By reference to the account with McCalmont Brs. & Co. there will be found in their hands a balance, to your credit, of £146.12.7. It will be necessary to draw on this sum for the amount of Elliott Brs' last bill amounting to £137.2, when there will be left a balance of £9.10.7 in London.

Hoping that my management of the business may meet your approval

I am, Yours very truly
John P. Marshall"

In 1931, after being in Tufts University for sixty years, the specimen was exchanged with the Museum of Comparative Zoology (MCZ) at nearby Harvard University. It was probably decided that it belonged in a museum rather than a teaching institution. The subsequent history of the fossil was related to the author in a letter from Frederick J. Collier at the MCZ in Harvard:

> "For whatever reason the specimen was held here [i.e. in Harvard] for some years until the post-cranial skeleton [the skeleton without the head] was loaned to the Cleveland Museum of Natural History. Whatever their plans were they never materialized and the headless, disarticulated skeleton moved on to the Houston Museum. A skull was copied from another specimen and the Harvard bone material was cleaned, repaired and missing parts were replaced. A 'whole' specimen is now on display at the Houston Museum.
>
> The Whitby specimen, type of *R. longirostris*, is [therefore] in two places. The skull, in poor condition owing to pyrite decay is here at the MCZ in the care of Mr Charles Schaff . . . The remainder is on display at the Houston Museum . . ."

The move to the Houston Museum was documented in some papers sent to the author by the current Curator of Paleontology, Dr Christopher Cunningham, including this letter from the Director of the Cleveland Museum of Natural Sciences, to Dr Tom Pulley, his counterpart in Houston, in June 1961:

"Dear Tom
Since I have not heard from you in some time, I assume that some sort of hitch has developed in your hopes to purchase our dinosaur.

I am writing to suggest something which you yourself once thought about. As you know, we have a mounted Plesiosaur skeleton loaned from Harvard. We are not using it, and there is little likelyhood that we ever will. With your Harvard connections I am sure that you could get the loan transferred to your museum. We would have to expect payment of shipping costs and crating since the animal is quite large and heavy. The base is 6 feet 4 inches wide and 17 feet 4 inches long. The specimen separates into four portions.

If this suggestion appeals to you, I will await further instructions.

Sincerely yours
William W. Scheele"

The museum publicised the successful exchange in their monthly newsletter.

*The Houston Museum of Natural Science*
*Museum News . . .* December 1961

A PLESIOSAUR IS ON ITS WAY TO HOUSTON

Our Museum is obtaining from the Museum of Comparative Zoology at Harvard University a completely mounted skeleton of a great Mesozoic aquatic reptile known as a Plesiosaur. This skeleton has been in the Cleveland Museum of Natural History, but

> is being transferred to our Museum on a permanent loan basis. It is a specimen of *Macroplata longirostris* collected near Whitby, Yorkshire, England from beds of lower Jurassic age . . .
>
> The skeleton that we are obtaining is of average size, as Plesiosaurs go, measuring 17 feet in length and 6 feet in width, although record size for the long-necked species is 50 feet. Space is now being cleared for this new exhibit, and we hope that it will be ready for public view within a month.

The entire fossil is still owned by Harvard University (specimen no. MCZ 1033), though the headless body is on permanent, i.e. thirty-year, loan to Houston Museum of Natural Science, where it is on display to the public in the Hall of Paleontology – 2,000 miles from its head and 8,000 miles from Whitby.

chapter twenty-two

# Yorkshire Ammonites

## SAINTS AND GEOLOGISTS:

### *Whitby and its snakestones*

The original abbey at Whitby was founded in 657 by Hilda, Abbess at Hartlepool, who was sent by King Oswy. The monastery was for men and women. Thanks to Hilda's leadership, Whitby Abbey and monastery became important enough for a great synod of the churches of the North to be held there in 664. Saint Hilda's monastery was destroyed by Viking invaders in 867.

After the Norman invasion in the eleventh century three men, Aldwin, Elfwy and Reinfrid, made pilgrimages from the south of England to the holy places of the North. Reinfrid stayed at Whitby and re-established the monastery, with an abbey church. The remains of this building exist only at ground level. The majestic ruin that stands on the cliff today is a thirteenth-century building. The Abbey was ruined and the monastery buildings destroyed in the sixteenth-century Dissolution.

The first History of Whitby was written in 1779, by Lionel Charlton. Despite its famous abbey, Whitby amounted to nothing much until 1600, when the alum trade began. Charlton's book begins:

> "The town of Whitby, or *Whitebay*, so called from the colour of the waves breaking against the rock on the outside of the harbour, was by no means considerable for any thing but its Abbey, situate on the east side of

*Whitby Abbey perched high above the town on the East Cliff. The arms of the breakwaters can be seen marking the entrance to the harbour.*

the harbour, till after the year 1600. Neither Tacitus, nor any writer who has recorded the affairs of the Romans in Britain, having taken notice of any such place, shews that it either then did not exist, or consisted at most of an insignificant hamlet or two on each side of the river Eske: And there having at no time been found in Whitby, or its neighbourhood, any castle, encampment, inscription, coin, medal, urn, causeway, or other remain, such as the Romans left wheresoever they came, is an incontestible proof that they neither had a station, nor ever resided there in any considerable numbers. The Romans were not well enough acquainted with the art of navigation to venture their ships into a river so inconsiderable as the Eske is at low

> water: They rather chose (as the example of Caesar, when he first made his landing good in Britain) to bring the transports, in which the soldiers were, along-side of some beach, or into an open bay, where after disembarking the troops, they might let them ride at anchor: or, in the case of bad weather, haul them up on dry land, out of the sea's way. And this seems to have been their practice in Dunsley-Bay, about two or three miles westward of Whitby: For there, if tradition and the general consent of antiquarians do not deceive us, they frequently landed their soldiers, and marched them up into the country, as occasion required. And to confirm this opinion, we find Dunus Sinus, or Dunsley-Bay, mentioned by Ptolemy, the Egyptian geographer, as a landing place they frequently used."

When Saint Hilda first came to Whitby, then just a collection of fishing huts, she found the place infested with snakes. Such was her power that, with a wave of her hand, she turned them all to stone. These snakestones were found everywhere in the rocks of the cliffs and on the beaches. They were so abundant that the town put them on its coat of arms.

*The coat of arms of Whitby Abbey and town showing ammonites as snakestones.*

"We have a lamentable instance of the gross impositions practized on the ignorant by the religious in former times, at two of their establishments on this coast – Whitby and Lindisfarn, where, to make it appear to the vulgar that their titular saints possessed the power of working miracles; St Hilda is said to have decapitated the snakes and converted them into stones of that form (now the Ammonites of the Lias); and St Cuthbert of Holy Island, is said, with his little hammer, to have forged the introchi [?] (of the Mountain Limestone) – so called St Cuthbert's Beads.

What a gross perversion of Nature.

Happily the tricks of those days are gone by, never to return, and happy may the poorest man think himself that, he can now live in the full enjoyment of the expanding powers of intellect."

William Smith, 4 May 1836, Newboro Cottage, Scarboro

When Martin Simpson set out to catalogue the fossils of the Whitby region, he first gave a short history of his subject.

"The Fossils of the Yorkshire Lias, and especially the Ammonites, had been objects of superstition, and poetic fiction, from mediaeval times, but respecting their true nature, no one ventured to express an opinion.

Whilst it was generally believed that the hills and valleys, and the whole face of nature, were the same as they ever had been, from the creation of the world, except some modification which it had suffered by a transient deluge, the forms of organized beings found deeply buried in the earth remained inexplicable. It was only about the middle of the last century that correct views respecting the changes which had taken place on the earth's surface were promulgated, and the origin of the noble, sublime, and popular science of Geology, now established upon indubitable evidence, can scarcely be placed earlier than the beginning of the present [century]; whilst the systematic study of Fossil Remains, or *Palaeontology*, as it has been named, is still more recent . . .

The publication [of Young's *History of Whitby*, 1817] immediately produced a general revolution in publick opinion respecting the fossil remains of the district, and excited great zeal for further discovery. There was, indeed, at this time, in Whitby, a strong desire after intellectual pursuits, not only amongst the learned, but amongst many whose circumstances in life were unfavourable to such pursuits. The cessation of a long and exhausting war, the energies aroused by that war, and the want of employment before the return of commercial prosperity, all had a tendency to intellectual pursuits, and, no doubt, contributed greatly to the establishment of Philosophical Institutions and Museums, which the great wealth and the national prosperity of the present era scarcely sustain . . .

The discovery of the remains of animals and plants buried deep in the earth, possessing all the essential characters of living beings, has given a great impulse to the study of natural history in our times; and it is chiefly on the evidence disclosed by fossil remains, that there has been constructed the new, popular, and sublime science of Geology."

Martin Simpson, 1855

An ammonite is an ammonoid. There are three groups of ammonoids, distinguished by the markings or suture lines on their shells. Goniatites have angular zigzag lines, ceratites have lines in the form of 'lobes', often with frills bursting out of the lobes; ammonites have complex frills.

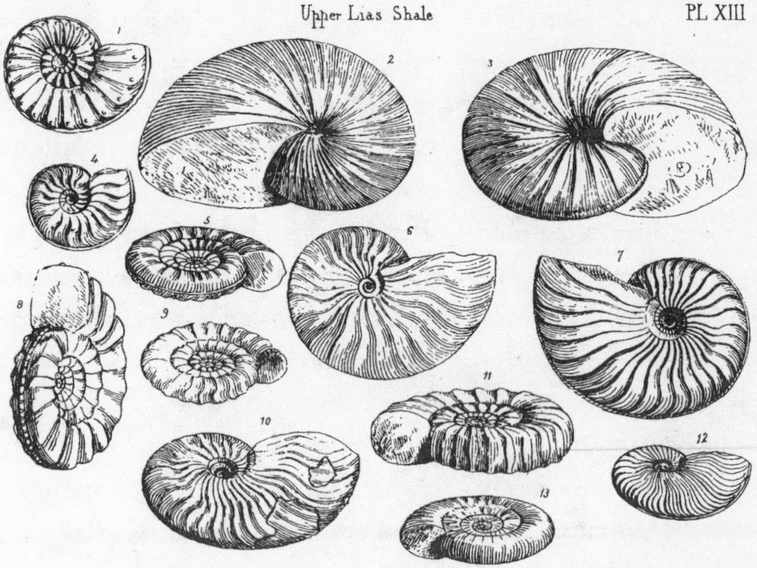


*Ammonites from the Upper Lias Shale of the Yorkshire Coast (from* Illustrations of the Geology of Yorkshire, *John Phillips, 1829, Plate XIII).*

Ammonoids are cephalopods, all of which have mouths surrounded by at least eight tentacles. Octopus and squid are also cephalopods, as are belemnites.

Cephalopods are, in turn, molluscs. All molluscs have a foot, made of muscle tissue, and adapted for different uses in different groups. Bivalves, gastropods and cephalopods are mollusc groups which are common as fossils.

Ammonoids have a flat coiled shell containg a number of chambers. The animal lives in the outermost chamber, and controls the buoyancy of the shell through a long tube, or siphuncle, which connects all of the chambers.

Ammonoids are used as zone fossils for the Mesozoic era in many parts of the world. The Mesozoic era lasted from about 250 million years ago to about 65 million years ago. Within the Mesozoic era are the following geological periods: Triassic, Jurassic, Cretaceous. Ammonoids became extinct at the end of the Mesozoic era.

A zone fossil is really the remains of one of a group of organisms whose members are quite bad at adapting to changes in their environment. They tend to die out regularly, but then new species of the same group arise to take their place. The geologist can trace the passing of time through the changing species in the group, and can cross-match different rock formations which are geographically separated – if they contain the same species of zone or index fossil, they should be of the same age.

To be really useful, each new species of zone fossil must spread widely and rapidly – the more widely and the more rapidly the better. Rock from many hundreds of miles apart can then be cross-matched for age. The ideal zone fossils are free-floating plants and animals that disperse through the seas and oceans.

The usefulness of ammonites and other ammonoids as zone fossils, and the concept of detailed biostratigraphy, was discovered, or perhaps invented, by Louis Hunton of Loftus near Whitby in the 1830s. Biostratigraphy is the science of dating rocks by the biological remains they contain.

## A LIFE'S WORK

### *Louis Hunton and the discovery of biostratigraphy*

Louis Hunton was born in 1814 at Hummersea House near Loftus, ten miles north-west of Whitby, where his father was head alum-maker at the Loftus and Boulby Alum Works. Louis died twenty-four years later of tuberculosis in Nîmes in the south of France. The whole of his life's work is reproduced on the next few pages. Although this amounts to no more than 2,000 words of text (a systematic list of fossils and a detailed section were published with his original paper), it easily carries the burden of being the work of a whole, if tragically short, lifetime.

In his one published paper, read to the Geological Society when he was 21 years old, Hunton not only reported an entirely new discovery, and thereby initiated a new and endlessly fruitful science; he also laid down the methods by which this science must be practised.

It has often been thought, mistakenly, that science is somehow governed by a set of restrictive rules, and that if we can only find out what those rules are, we will understand what science is. Hunton states the major rule of the science he is creating, not in order to follow some philosophic code, but simply to say, 'Do it this way, or you will not be finding out what you think you are finding out.'

Before Hunton, geologists, led by the Frenchman Alexandre Brongniart and the Englishman William Smith, had come to understand that certain fossils were consistently associated with certain rock formations. This remarkable realisation transformed fossils from important objects of interest in themselves, into crucial tools of geological investigation. But this new science of biostratigraphy remained a crude measure until Hunton discovered its full potential. He realised that even within rock formations there are certain fossils which are restricted to certain bands of rock. Thus a fossil may be very common within a narrow band,

and not be found anywhere outside it. This alone made biostratigraphy a much more precise science; but Hunton had more to say.

He specifically understood that it was fruitless to look for changes in the overall fauna within a rock formation – some species will remain unchanged and ubiquitous through all the strata from an entire geological period. Instead the geologist must concentrate on those groups of animals which have shown themselves unable to adapt, and which therefore are represented by different species at different times. He was fortunate that the extensive alum works in the region where he lived exposed many thousands of specimens. He recognised that for the rocks of the Yorkshire coast, as it turned out to be for the whole Mesozoic era, the crucial fossil group was the ammonites. At a stroke this freed geologists from the need to identify every fossil within a rock formation in order to understand its temporal relations with other strata. They could simply look for what are known as *zone fossils*. Hunton then stated, at the end of his paper, that in order to be useful to the geologist, fossils must not be gathered from the loose debris on the beach, but must always be collected *in situ*: a rule that every geologist that has come after has unconsciously followed.

Louis Hunton is a romantic figure. His precocious genius, his early death from tuberculosis (that most Romantic of diseases) in a Mediterranean city, fit the requirements of an age that was ending even as he came into the world. But he was a scientist, not a poet. His name is occasionally revived by historians, who wonder that he is unknown to geologists, and then lapses again into obscurity. But though his one paper may not be much read or cited, his understanding has been unknowingly inherited by every geologist of the past 160 years.

*"Remarks on a Section of the Upper Lias and Marlstone of Yorkshire, showing the limited vertical range of the Species of Ammonites, and other Testacea, with their value as Geological Tests.*

By LOUIS HUNTON, Esq.

Communicated by J. F. ROYLE, Esq., F.G.S.

[Read 25 May 1836]

On presenting a stratigraphical arrangement of the fossils in the lias formation on the coast of Yorkshire, I wish to offer the following remarks relative to their geological distribution.

The section refers to that part of the coast between Whitby and Redcar, in which the Loftus and Boulby Alum Works are situated, and was named by General Mudge 'Easington Heights'. Its greatest elevation is 681 feet. It includes the inferior with a part of the great oolite series; and the lias is better developed in this place than in any other part of the coast, the upper shale having a thickness of 200 feet, and the marlstone of 160; and there are about 150 of the lower shale also exposed.

With regard to the correctness and universality of the order given to the fossils, the position of no one species was fixed till after several had been extracted from the situation indicated in the section, as well as in various different localities from the Alum Hills, fronting the Vale of Thirsk, to the peak near Whitby; and in the majority of instances, from thousands of specimens afforded by the quarrying of the lias shale in the manufacture of alum. Though it would have been easy to have added many more species, the geological situation of which is determined, yet wherever the limits of distribution have not been clearly defined, they have been omitted.

*Upper Lias Shale*. – Commencing at the surface of this deposit, the first fossil present in any quantity, is the *Nucula ovum*; which, after descending a few feet and entering what is technically called the "hard or cement stone seam," occurs in great abundance; but below that bed it becomes gradually scarcer and smaller, so that after a descent of 60 feet, it is rarely met with, and I believe has never been found so low as the jet rock. The *Orbicula reflexa*, and the small, but delicate and beautiful *Plagiostoma pectedoideum* are associated with the Nucula, and are similarly distributed. It may be worthy of observation that the *Nucula ovum* is singularly characteristic of that portion of the upper shale used in the manufacture of alum, and that wherever the shale occurs without this shell, it has been found unfit for the remunerative production of the salt. This is owing to the nature of the shale, the upper part, where calcined, being richest in sulphate of ammonia; not, as might be supposed, from containing originally *more sulphur*, but a *less proportion*, the lower strata having such an excess of pyrites and bitumen, that the contents of a kiln made from such schistus, invariably overroasts. A large Pallustra, or Amphidesma, (the *Amphidesma donaciforme* of Phillips) is also plentiful in the upper part; but it is occasionally found in the lower, and a similar shell occurs in the Marlstone. But of all organic remains the Ammonites afford the most beautiful illustration of the subdivision of strata, for they appear to have been the least able, of all the Lias genera, to conform to a change of external circumstances. Of the species, most plentiful in the upper part, scarcely any are found in the lower; except a few doubtful individuals of *A. Communis* (the prevailing species in the "hard seam") which are now and then seen so low as the jet rock; and though *A. fimbriatus* and *A. heterophyllus* are found in that stratum, they are always dwarfish, seldom attaining three inches in diameter, whereas in the upper beds, they are met with twenty inches in diameter.

The portion of the series mentioned before as the "Jet Rock", is a very compact and highly bituminous bed, from 20 to 30 feet thick and 150 feet below the surface of the formation. It contains many flattened, pyritous nodules, which generally contain organic remains. The rock itself is also so very sulphureous, that when a small heap of it was calcined at the Loftus works some time ago, the melted bitumen and sulphur flowed in flaming streams. The whole of the jet used in the manufacture of ornaments, is extracted from this stratum; for when occasionally met with higher up, it is like coal in its fracture, and too brittle to work well; and in the inferior strata, as the marlstone, it forms merely a thin coat, enveloping a lump of fossil wood. The seams are sometimes 20 feet long by 6 feet broad, but they are often smaller, and their greatest thickness is 3 inches.

The Jet Rock is also peculiar for containing the remains of Teleosaurus; for though Ichthyosauri and Plesiosauri appear pretty equally distributed through the upper shale, this (fluviatile?) reptile is rarely found but in, or immediately above, the Jet Rock. Its bones are generally scattered in separate nodules, the upper jaw being scarcely ever united to the lower. With the exception of the specimen in the Whitby Museum, I am not aware of any instance of this animal having been found entire, whereas the other saurians seldom want more than the head or paddles.

On inspection of the section, it will be perceived, that the Jet Rock has its peculiar suite of Ammonites, and here, as was before observed, *A. heterophyllus* and *A. fimbriatus* make their first appearance, being about the size of a half-crown, while in the "hard seam" they attain the enormous magnitude of 1 and 2 feet in diameter. The *Ammonites Crassulus*, (*n.s.*) common to the whole upper lias, on the other hand, has its greatest development in the Jet Rock, gradually diminishing from 1 ½ inches to ½ an inch in the overlying beds. An interesting, though perhaps it may appear trifling, circumstance attending this fossil is, that as it decreases in size it becomes more numerous, many hundreds being then found together.

The lowest bed of the upper shale is hard, compact and sandy, and is singular for its great barrenness of fossils, occurring as it does immediately beneath one so prolific as the Jet Rock. This scarcity, combined with the arenaceous nature of the stratum, may perhaps afford some hints for elucidating, hereafter, the anomaly of fluviatile reptiles being mixed up with pelagian shells and fishes. It is a plausible speculation, that during the formation of the upper lias, the bottom of the sea in which it was deposited, was gradually settling. This conjecture is supported, by the greater prevalence of vegetable matter and fluviatile (?) reptiles in the lower part, and the increase in number of the larger species of Ammonites, Nautili and Belemnites in the upper. This conjecture is further supported by the mineralogical nature of the schistus, silex being more abundant in the inferior parts, and alumina in the superior, an arrangement which would naturally take place, if the coast during that epoch gradually sunk, and the land, the detritus of which afforded the material for the lias, as gradually receded; the finer argillaceous sediment being consequently carried into the deeper and more and more distant parts.

*Marlstone*. – On entering this series, we at once perceive a striking contrast to the upper lias, for independently of the decided mineralogical difference, not only the species and genera, but the *order* of Testacea have undergone a total revolution. Instead of a preponderance of Cephalopods and a scarcity of bivalves, we now have an overwhelming majority of the latter, some of the beds being almost constituted of Cardia and Gryphytes, which have also an extensive vertical distribution. *Avicula Cygnipes, A. inaequivalvis* and *Pulastra antiqua* are equally abundant in the whole series; but other shells, as *Cardium truncatum*, occur more plentifully in the inferior strata, whilst *Terebratula triplicata* and *T. trilineata* crowd the superior.

The species of Ammonites, though few in number, are, however, highly characteristic; thus we find *A. vittatus* about the centre of the series, confined to a very small range, associated in nodules with the *Cardium multicostatum*, *Turbo undulatus*, and *Pecten planus*; but the two latter occur in other parts of the formation. The *A. maculatus*, is constantly found at the junction of the *marlstone* with the *lower lias*, which here pass so gradually into one another, that it is impossible to determine where the sandstones end and the blue shale begins. I have long sought for *A. maculatus* in the upper and central portions of the *marlstone*, but have never found it many feet above the junction beds; and though this and other Ammonites from unequal geographical distribution, may be more abundant in one place than in another (*A. maculata* is in greatest number at Staithes, *A. Hawskerensis* at Hawskerbottoms), yet they constantly maintain an invariable relative position. *A. Hawskerensis* I believe to be also a junction Ammonite, but from its rarity in other situations, than that mentioned, it is not entered in the table. The above description may not, in some instances, exactly accord with previous statements, but one great source of error has hitherto been, the collecting of specimens from the debris of the whole formation, accumulated at the foot of cliffs or other similar situations, where they have long laid, and the inferring of their position from the nature of the matrix. A strong temptation to this method arises from the facility it affords of obtaining finer fossils than can possibly be procured from newly quarried nodules, the hard nature of which renders it very difficult to extract the fossil entire, without long exposure to air and moisture."

chapter twenty-three

# The Tenth Reptile

## *Human huskies – the Manchester plesiosaur*

The most recent great fossil reptile find on the Yorkshire coast was made in 1960. Members of Manchester University Geology Department found a skeleton of *Rhomaleosaurus longirostris* which turned out to be four metres long. Fred Broadhurst and Louis Duffy explained how the fossil was discovered and extracted. It is now on permanent display in the foyer of the Geology Department at Manchester.

> "The specimen was found between Old Peak and Blea Wyke Point, south-east of Robin Hood's Bay, in the *Hildoceras bifrons* Zone of the Alum Shale Series of the Upper Lias. The tip of the snout was discovered just projecting above the more or less horizontal (though undulating) surface of the wave-washed platform of shales, and after preliminary excavation and removal of skull, neck, pectoral girdle and one paddle bone (humerus), an expedition was mounted to recover the greater and more deeply buried part of the specimen . . .
>
> Half a dozen research students from the Geology Department of Manchester University worked under the direction of one of us (FMB)for two days in appalling weather conditions with driving sleet and hail. The shale matrix adhered strongly to the bone of the skeleton so that blocks of shale containing sections of the skeleton had to be collected. Drawings were made on the site to facilitate reconstruction . . .

The cliffs at this part of the Yorkshire coast reach a great height and the material collected had to be carried up to the nearest point of access by a vehicle, at approximately five hundred feet above sea-level. The route up the cliff lay first over a boulder field on the shore itself and then across rough ground to a narrow, steep and winding path. Specimens had to be man-handled as far as the path and were then loaded, in turn, on to a section of ladder (found washed up by the sea) which served as a sledge. Every member of the expedition was required to form a part of a human 'husky dog' team. A series of intermediate depots was organized to provide resting points on the way up.

Cleaning and preparation of the specimen have presented some interesting problems. The skull, neck, paddles and tail were enclosed by a relatively soft shale which could readily be removed mechanically, but the concretion developed in the region of the trunk presented a major problem. Chemical attack on the carbonate proved unacceptable because the bone present was attacked as readily as was the matrix by all the reagents tried. Fortunately the problem was solved for us by Messrs. J. and H. Patteson, stonemasons of Manchester, thanks to their manager Mr George Mills. Bench space was provided for us at Patteson's workshops and we removed the bulk of the matrix from above the rib cage and from the vicinity of the vertebral column with pneumatic chisels. Removal of the last remaining matrix from the surface of the vertebrae proved a long painstaking job and was done with ¼-inch chisels and very light hammers.

A second problem in connexion with preparation of the specimen resulted from recognition in it of 'pyrite disease', i.e. the production of corrosive acid due to the oxidation of pyrite, partly by bacterial action, within the specimen. Specimens not associated with soft shale matrix were treated overnight with a 50% solution of concentrated Savlon (ICI) and then coated with concentrated Savlon. In addition to arresting the effects of the 'pyrite disease' this treatment produced an unforeseen benefit in that the detergent power of the Savlon loosened and removed shaly material present in crevices on the surface of the bones.

The specimen now on display has a number of interesting features. The skull, for instance, has been detached from the neck and it is tempting to point to this severance as the cause of death. But further scrutiny of the skeleton reveals that various components were clearly moved about after death, possibly by water currents, possibly by scavengers, so that the neck break may itself be post mortem."

Frederick M. Broadhurst and Louis Duffy, 1970

chapter twenty-four

# A Museum at Scarborough

## *William Smith and the Rotunda*

In cramped laboratories and unsatisfactory lodgings, scientists in the early nineteenth century dreamed, it seems, of museums. William Smith did, and in 1804 or 1805 he scribbled his thoughts into his notebook:

> "Plans for fitting up a Museum
>
> Cabinet of Maps
> Do. of Sections of Strata
> Do. Arranged Fossils in Strata
>
> These to show them to the best advantage should be put upon Screens which will fold up or stand in the middle of a Room as occasion may require –
>
> Or the Maps and Sections may be pasted upon stiff paper or folding pasteboard and suspended in square cases under the ceiling and pull down and up by endless Lines &c.
>
> Two lines or cords running through two sets of Rings at each end of the Map one on the front and the other at the back of the Map would make them all fold up regularly into the square cases.
>
> Thin lines must be both drawn alike. The Boxes or Cases for these large Maps and Sections should hang upon Pulies so that they may be lowered at pleasure in order to examine any part of the Maps or Sections.
>
> All the Fossils arranged round a large room (with a

> Skielight) and the Maps & Sections suspended from above with a large drawing table in the middle seems to be the best way of fitting up a Mineralogical Museum."

In 1820, after many wanderings and misfortunes, including a spell in a debtors' prison, William Smith found a haven from the world in the seaside town of Scarborough. There he found a group of enthusiastic amateur geologists, eager to welcome him and learn from his experience. He was to return in 1824 and make Scarborough his home for the rest of his life.

> "I went to Scarboro' under distressing circumstances – unknown to anyone in the place but through the medical men . . . I there gained the best acquaintance and as my talent was always of use either in Town or Country, on the Coast or Inland, on the surface or to the greatest depths in the Earth I generaly found something more or less to do wherever I took my work . . .
>
> Everyone is very fond of talking on Geology; which makes it a good introduction to business, and as a geologist fully skilled in the stratification must know more of the Country than those who live in its midst [he] cannot fail to be noticed."

On his first visit in 1820 Smith put his idea of building a museum to the gentlemen naturalists of Scarborough. They were keen but there was not enough money to proceed. Five years later he tried again, and this time the money was found. Plans were drawn up by the York architect Richard Sharp, under instruction from Smith. He probably got the idea for the design from the Leverian Museum, a famous collection of fossils displayed in a building known as the Rotunda, near Blackfriars Bridge. The Leverian Museum closed in 1806. Smith and his backers decided the museum should be of 'such magnitude and character, as might comprise the whole of the collection in one room, to hold meetings at which scientific communications will be received and to establish a library.'

The foundation stone for the Scarborough Museum was laid in 1828, bearing this inscription:

THIS BUILDING, ERECTED FOR A MUSEUM
BY SUBSCRIPTION OF THE MEMBERS OF THE
SCARBOROUGH PHILOSOPHICAL SOCIETY,
WAS BEGUN APRIL 9, 1828

*THE PRINCIPAL PROJECTORS WERE*
SIR JOHN V. B. JOHNSTONE, BART., PRESIDENT
THOMAS DUESBERY, ESQ.,
ROBERT TINDALL, JUN. ESQ., CHAIRMAN OF THE
BUILDING COMMITTEE;
JOHN DUNN, ESQ., SECRETARY;
WILLIAM SMITH, ESQ., GEOLOGIST;
MR BEAN AND MR WILLIAMSON, NATURALISTS.

Opposite and on pages 316–7: *Architectural drawings prepared by R. H. Sharp of York, under the direction of William Smith, for a museum at Scarborough. The elevation, section and plan are shown, together with the design for enlarging the museum which was carried out in the 1860s.*

ELEVATION.

SCARBRO' MUSEUM.

Designed by R H Sharp Arch^t York.

## SECTION.

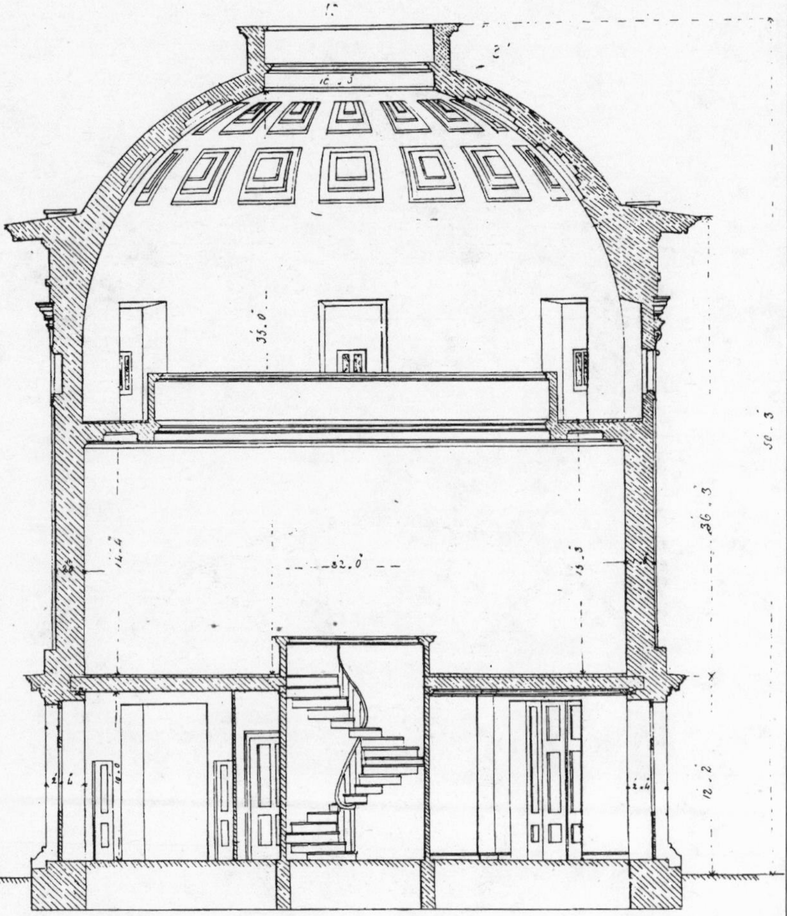


## SCARBRO' MUSEUM.

Designed by R. H. Sharp, Archt. York.

DESIGN FOR ENLARGING THE MUSEUM.

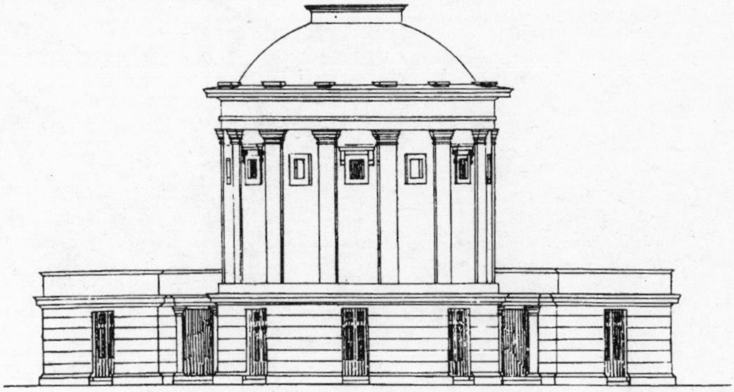

ELEVATION

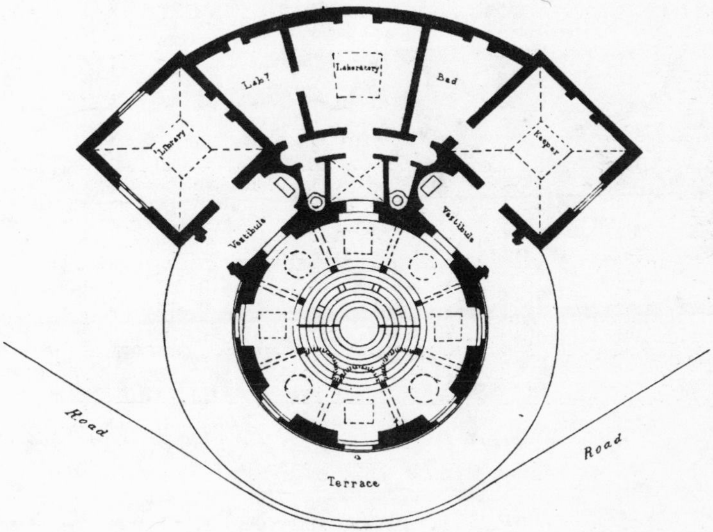


PLAN.

SCARBRO' MUSEUM.

The first proceedings of the Scarborough Philosophical Society were held with a Public Dinner to celebrate the opening of the Scarborough Museum on 31 August 1829. The report begins with a description of the new museum, before a brief summary of the dinner:

> "Amongst the numerous interesting objects with which the place abounds, the new Museum now stands most prominent and attractive. It is situate to the south of the bridge, on an ascending piece of ground, and is seen from the sands rising majestically above that beautiful erection.
>
> The Museum is a rotunda of the Roman Doric order, 37 feet 6 inches in external diameter, and 50 feet high. The basement contains, pro tempore, the library, keeper's room, and laboratory. When sufficient funds are obtained, it is proposed, to place these accessories in wings radiating from the central building, which will then be entirely used as a Museum. The principal room is 35 feet high, and is lighted by a central eye or opening. The beautiful Hackness stone, the munificent gift of Sir John B. Johnstone, Bart., has been employed in this building. The fossils, which are very numerous, are arranged on sloping shelves, in the order of their strata, – shewing at one view, the whole series of the kingdom. A horizontal shelf below sustains the generic arrangement of fossil shells. Amongst the collection of fossils, which is one of the most perfect in England, are two admirable collections of local fossils, one purchased of Mr Williamson, and the other presented to the Society by Mr Duesbury, being the valuable collection of the late Mr Hinderwell. The birds and animals are

placed above the geological arrangement; so that every part of the Museum can be seen at once. The whole expense of the building, fitting up, &c., will be about £1400., of which £1100. has been raised. The remainder, if not contributed by the liberality of the friends of science, must be raised by loan, on interest. A donation of £25 constitutes a proprietor's share, which is transferable at all times by will or sale, and confers a perpetual right of admission to the family of the owner. A donation of £5 confers the same perpetual right to the families of strangers. The terms of admission to casual visitors are reasonable; and the receipts from this source will go to pay the keeper, Mr Williamson, who is always there to explain the geology of the district.

The building has been erected from designs by Mr R. H. Sharp, architect, of York; on whom it reflects much credit.

## PUBLIC DINNER

Monday, Aug 31st, [1829] being fixed upon for the opening of the Museum, it was determined, by the Committee of the Institution, that a public dinner in commemoration of the event should take place at Houson's hotel. At four o'clock on that day, accordingly, sixty-two gentlemen sat down to a sumptuous dinner, – the table being spread with every delicacy of the season; with a fine dessert, and excellent wines. Amongst the profusion of confectionaries &c which decorated the festive board, was a model of the Museum and surrounding garden, furnished by Mr. Gray,

> confectioner; it stood between two and three feet high; was three feet diameter at the base; and was much admired."

The 'Rotunda' Museum still stands near Scarborough seafront, one of the oldest surviving purpose-built museums in existence. The geological collection was moved to Scarborough's Natural History Museum in the 1950s, and the wall-cases lining the circular room are now filled with various memorabilia of Scarborough's past. The geological section of the Yorkshire coast, drawn by John Phillips around the inside of the room, remains, though; as does the slight lifting of the heart that comes on emerging from the spiral staircase into the exact centre of Sharp's exquisite domed chamber.

chapter twenty-five

# Ice, Water and Landscape

## SLACKS, SWANGS AND RIGGS

*Landforms on the North York Moors*

Percy Fry Kendall's work on the ice age landscape of north-east Yorkshire, first published in 1902, showed how the present land bears the marks of its extraordinary past. Evidence of a great system of lakes, linked by ever-changing drainage channels, is found almost everywhere on the North York Moors, and is preserved in the names of many of the landforms – this is a small selection. Kendall identified three groups of lakes.

### The Eskdale Lakes

*Eskdale* (680084 to 900110) The valley of the River Esk now runs from Castleton to the sea at Whitby. In the last ice age the western part of Eskdale was the centre of a great lake.

*Lealholm Rigg* (752083) Ridge created by the terminal moraine left at the edge of the Scandinavian ice sheet. The wall of ice which existed here for thousands of years formed the eastern limit of Lake Eskdale.

*Danby Dale, Kildale, Sleddale, Commondale, Baysdale, Westerdale, Little Fryup, Great Fryup, Glaisdale, Egton Grange* (within the square 620040 north-east to 790090) Tributary valleys of the Esk which once formed part of Lake Eskdale.

*Vale of Mowbray* The low area to the west of the escarpment which marks the edge of the North York Moors. An ice sheet pushed down the

Vale in the last ice age, but did not get over the escarpment on to the high moors.

*Murk Esk (Muir Esk)* (823037) The stream which follows, in reverse, the route of the great overflow channel from Lake Eskdale to the south.

*Wheeldale* (814985) Isolated valley which was the southernmost outlier of Lake Eskdale.

*Murk Mire Moor* (800020) High land to the west of Goathland valley.

*Two Howes Rigg* (825994) Spur of high ground separating Wheeldale and the valley of the Murk Esk.

*Nelly Hay Force (Nelly Ayre Foss)* (814997) Waterfall at the furthest point of ice sheet penetration up Wheeldale, marked by a bank of moraine.

*Lady Bridge Slack* (802024) U-shaped trench across Murk Mire Moor which marks the old channel of the waters draining from Lake Eskdale towards Newtondale.

*Purse Dyke Slack* (803017) Trench or channel which carried the Eskdale waters further south.

*Moss Slack* (827002) Abandoned channel which formed a link between the lakes in Wheeldale and the valley of the Eller Beck to the west.

*Moss Swang* (808025) Escarpment running along the hillside above Struntry Carr. This massive overflow channel was bounded on one side by the ice sheet; on the other side the waters gouged a steep bank out of the side of Murk Mire Moor.

*Randay Mere* (811019) Abandoned glacial overflow, which has been dammed at each end and turned into a reservoir.

*Newtondale* (853975 to 804853) Deep gorge running for twelve miles through the plateau of the moors. Created by the outflow of waters from Lake Eskdale to Lake Pickering, this is the finest example of a glacial-lake overflow channel in England.

*Fen Bog* (853975) Low boggy ground at the present watershed of the great Newtondale overflow channel. Eller Beck flows to the north, Pickering Beck to the south.

*Crunkly Gill* (755072) Narrow channel cut through the moraine west of Lealhom. The River Esk runs through a gorge at Crunkly Gill.

*Wild Slack* (753069) Abandoned overflow channel on the high part of Lealhom moraine above Crunkly Gill.

*Sunny Brake Slack* (772063) This old glacial water channel, known as an 'in and out', loops away from Eskdale, round a prominent hill and back into the main valley. Its course was probably around a protruding lobe of ice. The channel would have been abandoned once the ice withdrew, and allowed the original valley of the Esk to be used as the main channel.

*High Brook Rigg* (773063) High ground overlooking Sunny Brake Slack.

*Sleights Moor* (855045) High ground to the east of Newtondale on the south bank of the Esk.

*Ewe Crag Slack* (692112) Huge overflow channel on the north side of Eskdale.

*Mullion (Mallyan) Spout* (824010) Once Lake Eskdale drained away to the south, streams resumed their natural drainage northwards. Wheeldale Beck meandered over the plain before cutting a spectacular gorge west of Goathland, with its famous waterfall.

## Northern Lakes

*Bold Venture* (605130) Ironstone quarry on the site of a small glacial lake and overflow channel. Ice age lakes and water channels were created on the northern margin of the moors, in small spaces between the ice sheet and the high ground.

*Slapewath* (638150) Overflow channels run from the escarpment above Slapewath, east to Moorsholm Rigg, eight miles away. 'Wath' is a Scandinavian word for ford; 'Slape' is used locally for slippery. Being a geologist, Kendall suggested that it is the fine-grained lias stone that makes the ford at Slapewath slippery.

*Stonegate Valley* (775095) As the ice sheet retreated this valley opened up to drain a section of the northern moorland into the Esk.

*Barton Howl* (803084) Steep valley formed by overflow channel from a small glacial lake at the top of the watershed.

## Coastal Lakes

*Robin Hood's Bay* (955045) The curve of the bay is due to the erosion of a dome-shaped structure in the underlying rock. The flat area of ground above the shoreline is glacial mud, as it was left by the retreating ice sheet.

*Mill Beck* (950038) Cuts down through glacial mud and underlying Lias rock; runs to the shoreline at Boggle Hole.

*Stoup Beck* (955033) Stream cutting straight through glacial mud down to the shoreline at Robin Hood's Bay.

*Lilla Howe* (890986) and *Blea Hill Rigg* (903005) High points of land which remained ice free throughout the last ice age.

*Biller Howe Dale Slack* (906018) Channel cut by glacial waters draining south into Jugger Howe Beck. This feature manages to carry three generic descriptive terms; Howe = barrow or low hill; Dale = valley; Slack = abandoned drainage channel. Biller Howe Dale Slack is therefore the abandoned channel that runs into the valley that runs near the hill at Biller. Biller Howe is now the name of a farmhouse.

*Sneaton Low Moor* (910035) Area of moorland cut by outflow channels as the coastal ice sheet retreated.

*Far Middle Slack (Sike)* (910033) and *Nigh Middle Slack (Sike)* (912038) Old outflow channels crossing Sneaton Low Moor. Far Middle Sike has a small stream running into Kirk Moor Beck, which explains the term 'sike' being preferred to 'slack'.

*Jugger Howe Beck* (940988) Stream flowing through the massive overflow channel of Jugger Howe Slack, which drained a series of small lakes in the hills above Robin Hood's Bay. Kendall describes it as 'probably the most curiously-placed stream-course in the country', because the present stream follows the same valley as a previous one, but on an entirely different route.

*Whisperdales* (960933) One of a series of dales which are cut back into the high ground of Silpho Moor by springs. These dales feed into Hackness, which was the site of an ice age lake.

*Forge Valley* (985865) Deep wooded gorge formed by an overflow from Lake Hackness to Lake Pickering. Now carries the waters of the River Derwent.

*Hagworm Hill* (005877) High point of ground on Seamer Moor, inland from and above Scarborough. This and other hillocks – Baron Albert's Tower, Seamer Beacon – are made of the terminal moraine of the great Scandinavian ice sheet, and mark its furthest progress inland.

*Marra-mat* (*Waydale* or *Weydale*) (016856) Abandoned outflow channel related to the retreat of an ice lobe which penetrated inland for about five miles up the present Vale of Pickering, forming the eastern margin of Lake Pickering.

## FROM UNDER THE ICE:

*In the footsteps of a scientist*

When it came to my turn, the beginning did not appear. Where was it after all? In me and my entry into it, or perhaps in Kendall's, or even in its own? And where would that be? 'Water,' I blurted out, then stopped.

The face of the old man came nearer and said, 'Take your time, you are among friends here.' I thought, Who wouldn't say that?

Too much drink had been taken, that was for sure. The beer and the whisky and the heat from the fire. My mind was bubbling with things that could be said. Getting from thought to articulation was the problem – the trick that had been lost.

I was thinking, Is it a story anyway? and, Aren't there any others to tell?

But they had all told theirs. I owed them one, and there was only this one.

'It's all to do with water and ice,' I said, and the bottle was passed again. Someone to my left filled my glass. It went quiet enough to hear the snow fall against the windows. 'It begins right here, about 20,000 years ago.'

'Before your time, Bill,' someone said, and other voices hushed him into silence.

'Or a hundred years ago.' Yes, I thought, Exactly a hundred years ago. 'Or maybe last summer.'

I reached down into a bag for the paper I always kept with me. 'A system of glacier-lakes in the Cleveland Hills' by Percy Fry Kendall, published in the *Quarterly Journal of the Geological Society*, August 1902. A scientific paper in a scientific journal, discovered by a roundabout of references, in the library of the Geological Society of London, on a sweltering afternoon last August. And though every paper that is published has its audience, this paper was out of the ordinary – quoted by every geologist who has worked on the ice-made landscape of the northern continents since Kendall.

What could I begin to describe? What story of Kendall, or of me, or of the land where we sat, 400 feet below the surface of the lake that once filled this valley? But I was thinking first of the library. Of the remembered sense of sitting in the middle of that three-storeyed room, and the physical presence of its books and journals. Of how you could so easily step out of time in a place like that. And how, by connecting yourself into that tower of accumulated information, you gained a tiny piece of infinity and lost something much more difficult to define. I remembered how it felt, as if you were watching helpless and fascinated as a part of you floated away. Seduced into living for ever in the bank of knowledge – discrete, learned, well-referenced and meticulously classified.

You did not need to be 'unscientific' to feel the potency of that lined space. Climbing to the upper gallery with a scribbled shelf number and walking before thousands of volumes bearing thousands of papers. How many of them are ever looked at, even to be discounted, ticked off some diligent researcher's list? Am I the only one who is always surprised (and joyed) to find that this paper is exactly where someone said it would be? And that secret pleasure of the worker in the obscure places of scientific endeavour, knowing that only a few – but there *have* been a few, to have none would be less piquant – have come to this shelf, to this volume, to this page. You use it, you replace it, and you leave no trace of your passing, you leave no mark in its margins. You are the reason for its existence, but it goes on without you. Someone else will come. Sooner or later.

But this high and cool library, would not, I could tell, go well in a low inn, hot and stuffy in its fireside bar, on a winter's night. The snowflakes were turning to icy water on the warm window panes. It was hard to tell, but it looked as if the watery snow had reached up to the level of the windowsill. It was impossible to imagine that it would not snow for ever. The red flickering faces were expectant. The drink, the warmth and the snow had relaxed them. There was nowhere to go, they had all told their stories, they weren't eager to bring the night to a close. I read the first few lines of Kendall's paper.

> 'It is now 8 years since I commenced observations upon the Glacial phenomena of the Cleveland area – using the term in its broadest sense as including the whole of the Jurassic mass of North-eastern Yorkshire, but no results were attained in the first five years.'

'It is a story of a man of great patience,' I said, 'and of a kind of curiosity. But not ordinary curiosity. This man walked the land above our heads a hundred years ago, with a . . . a *faith*, that there was something in the land. Something he could sense from its abnormalities, and from the signs and traces left on the land. So he walked over the land for year after year after year, measuring and observing and noting and marking maps. Accumulating evidence of something he was not yet sure of.'

And as I said this I wondered that we had so arranged things that a man might not just be able to do this – he was paid by our forebears, was he not, to do this? – but that he should feel that this was a thing to be done. I had no illusions about the sanctity of science; to me it was something we did because we could do it, and I was glad that we did. Glad that we thought it important, or at least desirable, that a man of intelligence and insight should be enabled to turn his gifts to such a contemplation. The case for science in the abstract, or as something useful, I had put aside for now. I had simply known that there was something in Kendall's work for me, but that I too must be patient, and work diligently to extract it. But, let me present no illusions, I had wanted reward for my patience. I read on:

> 'The investigation, however, of a tract of country on the western edge of the Vale of York familiarized me with many of the more patent signs of old glacier-lakes, especially the phenomena of abandoned overflows, and this experience enabled me to recognize the great Newton-Dale valley as belonging to this category. Following this clue backward, I discovered the great system of lakes and related phenomena now to be described.'

'You will all know, no doubt, that this country' – I nodded to the blackness visible above the watery snow outside the window – 'was once covered with ice. But you might well find yourselves in the wrong. That might not matter much to you. Who cares if you get a few details wrong; you can't be expected to know everything. No one can know everything.' I paused and sipped at my glass. I could see from their faces that they weren't sure if they were going to like what they were hearing. It did not sound like a proper story. But they were intrigued enough, or inert enough, to stay with it. 'But you can know more than you do, and if you live and walk and work and love this piece of land, then you might owe it to yourself to know more about it.' I found myself hurrying to say things that would make them believe that they should listen. 'Don't listen to the voices that say that knowledge destroys beauty, that the more you know of something, the less it speaks to your heart, that knowledge puts a filter in the way of feeling. Knowledge isn't like that. Knowledge makes you look again, and again, and again. It is a virtuous circle. Ladies and gentlemen, shall we enter the circle of knowledge?'

There were murmurs of consent, and the bottle came round again. As we drank, my thoughts flew momentarily to the high plateau 500 feet above us, and to the first time I walked across the North York Moors. I fell in love with them in that way that makes you want everyone you love to love them too. And so you take someone with you in your imagination, and show them and tell them about this place, this land.

On a gloomy day, in the dull part of the year, the high moors present the most desolate sight of any landscape in England. The endless undulating expanse of dark brown vegetation – burnt patches of heather alternating with dead bracken – broken only by the occasional clump of gorse or solemn hawthorn, seems to suck the colour from the world, so that the sky can only surrender to the general mood and present a canopy of deep grey. The topography is arranged against your expectations. The high part of the land is flat and uninhabited, while the lower parts are steep and fertile. On such a day you may sit on a stump of thorn in the central part of the upland, like Hardy on his everlasting heath, and see nothing of the world but the brooding moor. The only hint you have that other things may exist somewhere, is the deeper shadow in the landscape where a dale dips below the central arch. Otherwise all is sombre, and all is grandeur. It matters little that the moors are a human creation – burnt off regularly to allow heather shoots to grow for game birds. The moors seem to have endured for ever because their qualities are enduring. The frivolous coat of purple they get up in the high summer is startling, picturesque, and fleeting. It is in the darker days of the year that they show their natural profundity.

There were many such days in my travels over the moors. And though I first found myself a little afraid of the potency of the landscape, I soon began to value the sense of dramatic permanence that it invoked. To be on top of the world, alone on this wild heath, was a source of either desolation or invigoration. I came to see that the difference between these responses was a simple act of will. But the high moors are only part of this landscape. North and south from this flattened brown arch runs a series of green dales, the like of which you will not see anywhere in this world. Walking over the moorland, your mind floating above the monotonous lunar surface, you will not expect what is about to happen. Then, in a heart-stopping moment, your eye catches a flash of green and the land before you dives down. And you step forward and it dives further, and then further, until you are standing on the lip of the world. Suddenly, from a sterile moonscape you are gazing into a mosaic of pastures and trees and hedges and farmsteads and villages. From the dowdy and colourless heath you see the bright, verdant, hopeful confusion of human interaction with the natural world. Looking down into a dale on the North York Moors is to wonder Gulliver-like at the trouble that has been gone to, simply to provide something of such perfect beauty for your delight. If you are lucky – why should you not be lucky? – the sun will respond and shed a warm yellow light from behind your shoulder on to the land below. In that light fields and hills and farms seen across miles of open valley come rushing through the transparent air to present themselves, for your pleasure, as perfect miniatures not ten inches from your nose.

But then a few hours later or the next day you might see another, and you begin to realise that perfection is an altogether inadequate human construction. These valleys are deliciously unimaginable. Step over the brow of the rigg and stop. See Rosedale go on for ever, curving lazily to the south, showing a hundred and twenty different shades of green in its hundred and twenty stonewalled fields. You are engulfed, and you know that this is what you wanted to see the instant before you saw it. And yet you could not have summoned this from your imagination. In this the world is strangely, for so self-possessed a thing, beyond human understanding.

The sound of the clock chiming above the fire brought me back to the room and the table in the inn. I half noticed that the snow was now drifting or settling against the window. The level was creeping higher. I took a mouth of beer and then a sip of whisky, and then took up the story once again.

'Six months ago I started my pursuit of a man who came here one hundred years ago, in pursuit of what had happened here 20,000 years ago. I found this paper' – I placed my hand across the opening page – 'in a library two hundred miles away. It told me that there were great things to be found in the country where I lived. All I had to do was to learn to look at the country through the eyes of this man – Percy Kendall. There would be difficulties. Physical to some extent, but mainly mental. I am not, I confess, a methodical person, but to travel with Kendall, and to be brought to an understanding, a certain amount of method is necessary. And, like Kendall, it is necessary to walk for many miles and look for many hours before anything emerges from the land – if I had known how many hours I might never have begun.' I said this, though I knew well enough that I would have. 'I once heard someone say that every great footballer has a picture in his head of the layout of play at any time – a kind of map, so that he can move himself and the ball as if he is seeing everything from above. Kendall was like that too. He seemed to have an instinctive grasp of the layout of the landscape, at any time in its history. I knew I would have to learn some of that ability to travel with him.'

I saw the landlord begin to scratch his ear, and moved on.

'Let's go back 100 million years. The place where we are sitting is on the coast of a subtropical sea – sometimes a river delta forms, sometimes the sea pushes in and floods the shallow lowlands, but mostly there is swamp. Enormous reptiles swim through the shallow seas hunting for fish, while beautiful coiled ammonoids and squid-like creatures known as belemnites forage near the sea bottom. The river brings silt, the swamp forms mud and the shallow seas bring beach sands and limy sludges. All these are being buried as more sand, silt and mud is piled on top. Millions of shells of ammonites and belemnites and thousands of reptile skeletons fall to the sea floor. It is all being compressed and turned into huge beds of rock. And then, about 70 million years ago, the sea level rises, and the country is drowned in a crystal clear blue tropical sea. Great beds of chalk thousands of feet thick are formed at the bottom of the clear sea, pushing the sea-bed down as they pile up.'

The faces were intent now, though everyone had leaned further back in their chairs. Their minds were engaged, their bellies were full. Another episode in the Earth's history began.

'Then, about 60 million years ago, the sea level drops and the whole region is lifted up. The chalk and the sandstones and limestones and shales now stand above the ground towering above the sea. But even as they are being lifted, the process of erosion begins. Rivers pour down from the west, wearing away the soft chalk. Five thousand feet, a whole mile of rock, is stripped away, leaving the old beds of limestone and shale from the coastal zone standing on the surface, just as they do today.'

A brow opposite me furrowed, as if concerned that this was the sudden end of the story. I drank and glanced at the rising snow in the window frame, and went on.

'In the middle of all this, about 40 million years ago, some part of a continent crashes into another and parts of the land buckle. There is slight uplift of land right here.'

I drew a line in the beer on the table. 'An upward arch is formed stretching from Ingleby to Ravenscar. Streams drain to the north, creating Westerdale, Fryupdale, Glaisdale, Iburndale and the rest.' I made lines with the wet beer. 'And to the south making Rosedale, Farndale, Bransdale. And that' – I leaned back in my chair for a moment, looking at the table – 'is where we were at the start of the ice ages, two million years ago, where the story begins.'

The landlord appeared from round the bar with a tray of beer. 'Looks like it could be a long one,' he said, placing a pint in front of each of us.

'Only two million years to go,' said someone, but no one was leaving.

As I drank I wondered if Kendall had ever been in this same room, and with the lightheadedness of the beer, I felt sure he had been sitting where I was sitting, a century ago, convincing himself by convincing others. Making stories out of signs. Someone turned the tree trunk on the fire and the red heat hit us like a flat pan. As I glanced out of the window the heat and the swirling flakes of snow disoriented me momentarily. I felt myself swimming, literally swimming, upwards through the valleys I had walked, upwards from the stone houses in the bottom of Eskdale, through blue-white water, towards the light, looking down at the fields and hedges and houses, seen through a film of blue so pure in colour that I felt myself beginning to cry, and then swimming up to break the surface of the lake and enter a world of sky and water and ice.

Someone coughed and my eyes, watery from the heat, came back into the room, to the flames in the grate and the marks on the table. I put down my beer and pressed on.

'The moorlands' – I placed a beer mat in front of me – 'were peculiar: a little too low lying and not the right shape for glaciers, but high enough to be an obstacle to the ice sheets coming from the north. So one great ice sheet pushed down the Vale of York on the west.' I drew a path one side of the beer mat. 'And another pushed in from Scandinavia, across the North Sea, and down the east side.' I traced a line down the other side. 'What Kendall set out to do, my friends, and it took him eight years to do it, was to figure out what happened in the middle.'

They craned nearer to the table as if by looking more closely at the beer mat, they might see better the truth of what I was saying.

'Now there's two things you should know, that you might know already. The Vale of Pickering, down here on the southern edge of the moors, was a lake until recently – I mean a few thousand years ago. That is fairly common knowledge. But where we are now in Eskdale, was also a lake. Where we are sitting was under three or four hundred feet of water for thousands and thousands of years. And this lake, twenty miles long and hundreds of feet deep, had never been discovered before Kendall came along. You understand that Kendall found signs of a lake all over the place – Danby Brick and Tile Works has beautiful clays full of minutely thin laminations, 200 to the inch, that could only form at the bottoms of lakes. But it is not as simple as that. The landscape is full of signs from its 100-million-year history – the geologist, as well as reading the signs, must decide the time at which each was made. Once Kendall "discovered" Lake Eskdale, he had to figure out how it came to be there. He knew that whatever answer he came up with would have to explain all the peculiarities of these moors – of which, as we all know, there are many.'

I retrieved a photocopied page from Kendall's great book on *The Geology of Yorkshire* which I always kept in my pocket, and read the paragraph that had first begun to open my eyes to this landscape:

> 'The geological student will find nowhere in Yorkshire a more remarkable contrast of scenery than in following the few miles of railway from Whitby up to Castleton. From the neighbourhood of Sleights he sees from his carriage window a valley encumbered with deposits of boulder-clay, and the river Esk makes its way down to the sea through many mounds of tumbled drift, through rock gorges and under steep bluffs of Lias. West of the Lealholm cutting in the moraine all boulder-clay ceases; the valley opens out wide, with a flat but terraced floor over which the river courses with intricate meanders. It is patent to every eye that here was once a great lake.'

The matter-of-fact prose, and the presentation of a conclusion as an obvious fact, deliberately disguising the months and years of tedious measurement and calculation, were more of an inspiration to me than the succulent prose of a hundred guidebooks. From the corner of my eye I saw that the snow was up to the cross-bar, halfway up the window – and still falling.

'We sit on the great divide of this valley. To the east the Esk behaves like a young stream, tumbling through steep gorges, and to the west like an aged river, meandering across a flat valley. But it should be the other way round, because the river flows, does it not, from west to east? The answer lies here. We are on the eastern edge of Lake Eskdale. Kendall knew this, but he did not know what was keeping the lake in this valley – why didn't it just empty into the sea? You see everyone assumed that lakes were formed by water from melting ice sheets, usually by valleys being bunged up with debris carried by glaciers. But here the debris – that ridge just beyond the village' – I nodded in a westerly direction– 'is about 400 feet above sea level, but the lake surface was 750 feet, right up to the dale heads. So, what was keeping the water in?' My mouth felt dry with talking. I drank more beer. I glanced at another passage in Kendall's work, and smiled at the excitement of his own discovery, hardly concealed by the formality of his prose:

> 'Some of the earlier geologists who recognised the signs of the tenancy of these Cleveland valleys by lakes . . . failed to realise the full magnitude of the lake phenomena because they looked upon the moraine as the obstructing barrier to the waters, and did not attach sufficient importance to the fact that an ice-sheet hundreds of feet thick would be equally effective as a barrier. Without this ice-wall there could have been no such high-level lakes as we have contemplated. And this

> moraine at Lealholm gives us proof of the effectiveness of the ice as a barrier.'

'This was Kendall's stroke of genius. He saw from the ridge of mud and rocks at Lealholm, and all the boulder clay to the east, that this was the western edge of the great Scandinavian ice sheet. He realised that the water didn't come after the ice, it was there at the same time, and that the ice sheet was the dam that made the lake. All the water in Lake Eskdale was held at its eastern edge by a gigantic wall of ice 500 feet high. And there it stayed for thousands of years.' There was a silence – perhaps lack of interest, or, I wanted to believe, contemplation.

'And then the ice melted and all the water drained away?' suggested a voice on the other side of the fire.

'It might have done that, but Kendall needed that Eskdale lake water for something else. He had one major landform to explain – the finest of its type in the whole country, and the one that I believe was his real obsession.'

It was time for the whisky bottle to come past again, and I was surprised to see that my glass was empty, before it was quietly refilled – it was an easy way to drink more than you should. I wondered whether I was making any sense at all, or whether anyone could tell. It was past the time to worry about that.

I looked at another passage I had marked in Kendall's work:

> 'It will be convenient now to interrupt our account of the happenings in Eskdale and its contiguous valleys to examine Newtondale. This dale, which played so important a part in the drainage system, is in many ways the finest example of a glacial-lake over-flow channel in England. It trenches entirely through the

> great Oolitic plateau which forms the flank of the main Cleveland anticline with an excavation in some places of more than 250 feet deep.'

The snow was three parts of the way up the window now – as high as a man's head. There was not much sense in hurrying. If we were being buried under snow, then it was too late to do anything about it. If this dale was filling with snow, then so were all the others. It would not take long to restore this rutted landscape to a clean, even, white plane. I read another paragraph while the whisky bottle distracted my audience:

> 'The most striking feature of Newtondale, whether we view it as a whole upon the map or travel through it by train, is its curious winding course. Other streams which flow down from the Cleveland anticline . . . are fairly straight in course. Newtondale, though it cuts the same rocks, displays its difference of origin by its sinuous trench. Evidently the volume of water which cut it was sufficient to obey the laws of meandering, though on the gigantic scale. The banks of normal rivers meander within their valleys. Here the trench itself meanders as it cuts its way across the upland plain. Harmoniously its banks have the characteristics

Opposite above: *Percy Fry Kendall's original map of the glacier-lakes of the North York Moors, at the time of the last ice age.* Below: *Simplified sketch of Kendall's map, showing present coastline, ice fronts. lake shorelines, and the two great outflow channels at Newtondale Gorge and Forge Valley.*

THE
GLACIERS & GLACIER LAKES
OF THE
CLEVELAND AREA
LAKE PICKERING
GLACIER OF THE VALE OF YORK

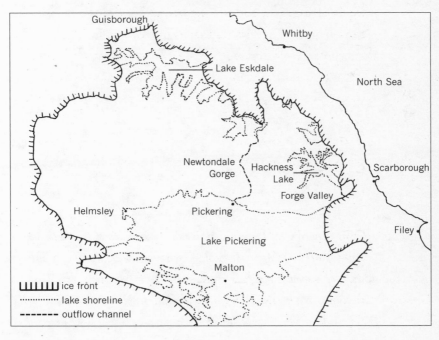

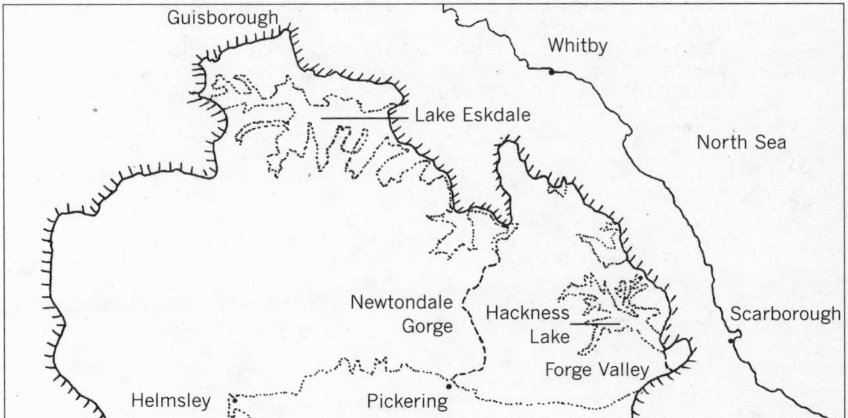
Guisborough
Whitby
Lake Eskdale
North Sea
Newtondale Gorge
Hackness Lake
Scarborough
Forge Valley
Helmsley
Pickering
Filey
Lake Pickering
Malton
ice front
lake shoreline
outflow channel

> shown by those of meandering streams. On the outer side of each curve the bank is usually an abrupt acclivity, while that of the inside of the curve is a gentler slope.'

I hurried on to the last part of my story.

'You've all seen the North York Moors Railway – probably been on it too. But have you ever looked at the valley it runs through? You should, because Newtondale is the jewel in the crown of this land. Get up on the high ground at Levisham Moor or Lockton High Moor or Two Howes Rigg, south of Goathland and walk over to the edge of the dale. Look up or down it, and then ask yourself if you've ever seen anything quite like it. Kendall knew he never had – except in the Alps. We all know that river valleys start off small and steep, and grow with the river. They end up broad and flat and straight, with the river meandering through the bottom. But look at Newtondale. The stream and its valley meander together, and the valley is a 250-foot-deep gorge more or less from beginning to end. It's as if someone had drawn a wavy line with a pencil through a block of butter. And the stream that's in it now is, to beg its pardon, pathetic. So, Kendall wondered at its beauty and asked himself, How did this great trench get here?'

I had found it a false game to follow anyone's thinking from what they had written, even in informal notes and letters. I liked to think of Kendall walking up Newtondale from Lake Pickering, becoming excited by its great walls, its exotic curves and its increasingly unlikely stream. I wanted him to have trodden those dozen miles with a growing expectation and a growing certainty of what he must find at the other end of this ribbon of space in the rock. I knew that the discourse of science had made certain demands of him, so that he must efface his personal experience from the official record; he must be there as a neutral objective observer. So I read in between the lines of his two

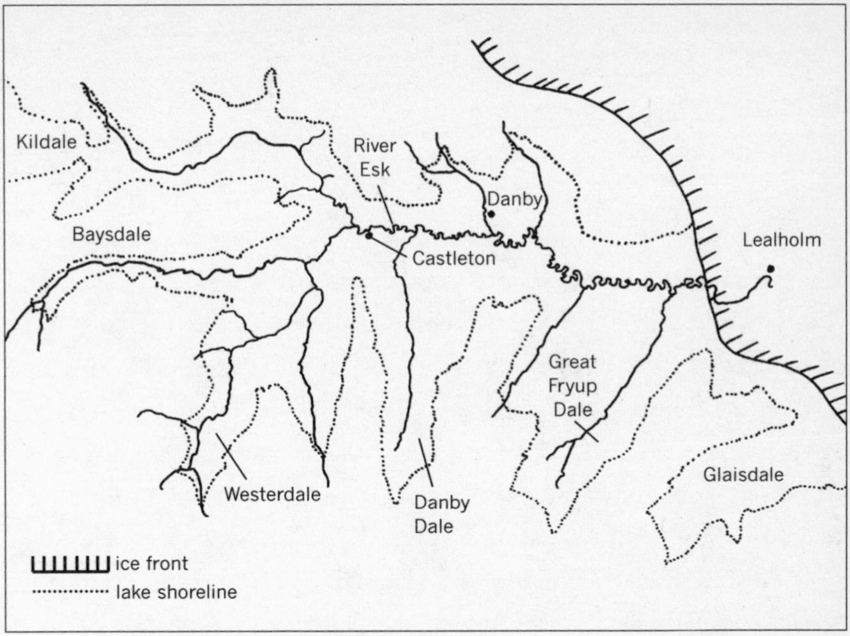


*The western part of present day Eskdale, showing the shoreline of Lake Eskdale at the time of the last ice age, about 20,000 years ago. The meandering course of the River Esk in the broad flat valley between Castleton and Lealholm gave Kendall a strong clue to the presence of the lake. A bank of moraine just west of Lealholm showed him the western edge of the Scandinavian ice sheet, which acted as a dam for Lake Eskdale.*

accounts – the dauntingly detailed 1902 paper, a hundred pages in length and eight years in the compiling, and the beautifully relaxed chapter in his 1924 *Geology of Yorkshire* – and made up my own story of Kendall and the ice lakes.

'Last home-made map,' I said, placing two beer mats six inches apart on the table. Faces leaned nearer.

'So here's Lake Eskdale.' I touched the one furthest from me. 'And here's Lake Pickering.' I placed a finger on the other. 'And here' – I dipped my finger into my beer and drew a sinuous curve between the mats – 'is Newtondale.'

The room was quite silent again. The window was now entirely covered. But the white covering was less like snow than a kind of half-frozen water, pale, reflecting the red firelight with a tinge of blue. That was our last measure. By the time I got upstairs I knew the surface would be past those low windows too. For the first time I began to feel cold, in spite of the fire and the whisky. And for the first time I wanted the story to end. I hurried on without any more hesitation.

'Kendall knew that such a valley could only be made by a huge volume of water cutting down through the plateau of rock. That way the water forces the valley to follow the stream, and the two meander together. This, he knew, was an overflow channel. What happened was this. The ice sheet was holding Lake Eskdale up at its eastern side. The other edges were held by the uplands on either side, with the western ice sheet in the Vale of York holding that end shut. About 20,000 years ago, at a point on the southern edge, just south of Goathland, natural erosion brought the land down level with the surface of the lake. Once there was a nick in the land, even just half an inch below the surface of the lake, the result was inevitable and spectacular. A lake twenty miles long and a few miles wide carries a lot of water. Just half an inch of depth of that area would be an immensely powerful torrent. And once the lake started flooding over the overflow it cut it down, so that more water flowed, which cut it down again. Down it went for over 200 feet, cutting the great gorge going south. Eventually it burst out here' – I pointed to the nearest mat – 'into Lake Pickering, and the two great bodies of water were joined. Kendall knew that if this had happened there would be a great washout of rock eroded from Newtondale – it all had to go somewhere. And sure enough he found it at Pickering – a huge fan of alluvial debris lying under the lake sediments. Once Lake Eskdale was drained, the water in Newtondale died to a trickle – a tiny stream in a great winding gorge. By that time the ice sheets had shrunk away, and the Esk could drain the lake to the sea.'

I leaned back into the comfort of my fireside chair. 'And that is the story of Eskdale, and the story of Kendall's discovery. There is more, of course. Once he figured out the lakes, he could show that there were abandoned overflows all over the moors . . .'

My eye was drawn to the window again. The pure blue-white was shining slightly. I was back on the surface of the lake, the sun reflecting blindingly off the still water surrounded by white walls of ice. I felt a shiver and then, quite strangely, a hand on my arm and a voice in my ear.

'Any more to drink, sir?'

It seemed to be from a place just outside the range of my perception. But then I felt the hand again, and the voice was much closer.

'Would you like another drink, sir, only I'm closing up the bar.' I felt my eyes opening.

'Oh.' I looked into the landlord's face: it took a moment to realise where I was. 'I think I must have dropped off for a minute.'

'I think you did, you probably had a long day. And that fire knocks it out of people. You're not the first to doze off in that seat.'

We were alone in the little back bar of the pub. The fire was dying down, and when I looked at the windows, there was no snow, just a blank black space. On my lap was a sheaf of papers. The top one was 'A System of Glacier-Lakes in the Cleveland Hills' by Percy Fry Kendall.

'So, another drink?'

'No, no thanks.' I looked at my watch, 'I'd better go to bed.'

In need of some air, I followed as the landlord showed his dogs into the car park for their late night wanderings. We were at the bottom of the great valley of the Esk with the riggs hundreds of feet above, black shapes against the dark grey sky. The night was balmy, with not a hint of snow or ice, or the frozen landscape that Kendall had re-created in his mind. As I looked up towards the high moors, I thought of the story that this land had told to him, and how he might have felt, being the first one to have seen it and read it; to be the first one to have collected the signs together, and made from them a cogent and persuasive history. I felt at least the echo of what his pleasure must have been, standing in the centre of it all with his words in my hand.

I thought about Percy Fry Kendall again as I lay in bed waiting for sleep. I thought about the pleasure of giving and taking. I had taken something from him, that was for sure. But I loved, in that curious fashion that humans have of judging their own actions, the way that I was giving to him, by paying such close attention to something he had taken such care to explain. I began to wonder too, whether this pleasure was caught up with a more selfish motive. Whether somehow I was foreshadowing the wish that I should leave something in the world, something that someone, some time would happen upon, and adopt, and take care to understand, and treasure.

# Chronology

Dates of the principal events mentioned in this book, and some related occurrences.

| | |
|---|---|
| 657 | Founding of an abbey at Whitby by Hilda, Abbess of Hartlepool. |
| *c.* 1600 | Beginnings of alum quarrying and manufacture on the Yorkshire coast. |
| 1746 | James Cook apprenticed as a seaman to John Walker of Whitby. |
| 1758 | Discovery of a fossil crocodile at Whitby by Chapman and Wooler, reported by them in the *Philosophical Transactions.* |
| 1768–71 | James Cook's first voyage of discovery on HMS *Endeavour.* |
| 1779 | *The History of Whitby* by Lionel Charlton is published. |
| 1787 | Miguel Rubin de Celis sends a sample of stone, now known to be a meteorite, from the Campo del Cielo in Argentina, to the Royal Society in London. |
| 1794 | William Smith confirms his theory of stratigraphy from the tower of York Minster. |
| 1795 | A meteorite is seen to fall in the grounds of Wold Cottage, home of Edward Topham. |
| 1802 | Publication of Edward Howard's report on fallen stones, which convinces the world that meteorites are extraterrestrial objects. |
| 1815–18 | William Smith sells his fossil collection to the British Museum to pay his debts. |

| | |
|---|---|
| 1816 | Publication of Kendall's *Descriptive Catalogue of the Minerals and Fossil Organic Remains of Scarborough and the Vicinity.* |
| 1819 | First ichthyosaur fossil discovered at Whitby, described by George Young in 1820. |
| 1820 | Edward Topham dies after erecting a monument on the spot where the Wold Cottage meteorite fell. |
| | William Smith sent to debtors' prison for ten weeks, makes his first visit to Scarborough. |
| 1821 | Discovery of thousands of bone fragments in a cave at Kirkdale. |
| 1822 | Publication of *A Geological Survey of the Yorkshire Coast* by George Young and John Bird, the first book on the geology of the entire coast. |
| 1823 | Publication of *Reliquiae Diluvianae,* William Buckland's ground-breaking book on the Kirkdale cave. |
| | Founding of the Yorkshire Philosophical Society in York, as a direct result of the collection of bones from the Kirkdale cave. |
| | Founding of Whitby Literary and Philosophical Society. |
| 1824 | Discovery of almost-perfect crocodile fossil at Whitby, now in Whitby Museum. |
| | William Smith gives a course of lectures in Scarborough, assisted by John Phillips. Although continuing to travel, Smith makes Scarborough his home. |
| 1825 | John Phillips made Keeper of York Museum. |
| 1826 | William Smith demonstrates the strata of the Yorkshire coast to a young Roderick Murchison. |
| 1827 | Founding of Scarborough Philosophical Society. |
| 1828 | William Smith finds financial security, employed as land steward on the estates at Hackness, near Scarborough. |
| | First account of giant plesiosaur specimen in |

Scarborough Museum, which has subsequently vanished.

1829 Publication of John Phillips's *Illustrations of the Geology of Yorkshire*, the first book on the geology of the region by a professional geologist.

Opening of the Rotunda Museum at Scarborough, designed by William Smith.

1831 Inaugural meeting of the British Association for the Advancement of Science held at York Museum.

William Smith named the 'Father of English Geology' and awarded the Geological Society's first Wollaston Medal by Adam Sedgwick at a ceremony in London.

1836 Louis Hunton's paper on the limited vertical range of ammonite fossils in the Yorkshire Lias is read to the Geological Society, and lays the ground for the science of biostratigraphy.

1841 Fossilised skeleton of massive plesiosaur found at Saltwick. Subsequently sold to Cambridge University after heated dispute with Whitby Museum.

1844 Plesiosaur skeleton unearthed at Kettleness alum works. Donated to York Museum, and then reclaimed by Marquis of Normanby. Specimen is now in Whitby Museum.

British Association meeting at York.

1848 Another enormous plesiosaur skeleton found at Kettleness alum works. This, the largest fossil found on the Yorkshire coast, is now in the National Museum of Ireland in Dublin.

1855 Martin Simpson publishes *Fossils of the Yorkshire Lias*.

1860s Huge ichthyosaur fossil found by jet workers at Hawsker. Now in Whitby Museum.

1876 John Potter Marshall buys a giant plesiosaur fossil in Whitby and transports it back to Boston, Massachusetts.

| | |
|---|---|
| | The body is now in Houston Museum, the head in Harvard. |
| 1880 | The alum works at Sandsend, the last to remain in operation on the Yorkshire coast, closes. |
| 1902 | Publication of Percy Fry Kendall's paper on the ice age lakes of Cleveland. |
| 1924 | Publication of Kendall and Wroot's *The Geology of Yorkshire.* |
| 1931 | Opening of the new Whitby Museum in Pannett Park. |
| 1953 | Publication of Hemingway's new theory of the Whitby fault. |
| 1960 | Plesiosaur fossil found at Old Peak, south of Robin Hood's Bay. The specimen is now in the geology department of Manchester University. |
| 1996-7 | National Lottery funds donated for cleaning and restoration of the internationally-important fossil reptile collection in Whitby Museum. |

# Notes and Sources

Some books and papers were used as sources for a number of chapters. The 1984 paper by Benton and Taylor, a brilliant combination of historical research and geological fieldwork, was the starting point for my own researches into each of the 'reptile tales' in the book. Kendall and Wroot's *The Geology of Yorkshire* contains fascinating diversions into the history of geology and geologists in Yorkshire, and much else besides. Browne's book is a delightful account of the history of the Whitby Literary and Philosophical Society and its various museums, while George Young's works are both meticulous and wondrously informative.

Benton, M. J. and Taylor, M. A. (1984) 'Marine reptiles from the Upper Lias (Lower Toarcian, Lower Jurassic) of the Yorkshire coast', *Proceedings of the Yorkshire Geological Society*, vol. 44, part 4, pp. 399–429

Browne, H. B. (1946) *Chapters Of Whitby History, 1823–1946*, Brown & Sons, Whitby

Kendall, Percy Fry and Wroot, Herbert E. (1924) *The Geology of Yorkshire*, 2 vols., P. F. Kendall (self-published), Leeds

Young, George (1817) *A History of Whitby*, Clark & Medd, Whitby

Young, George and Bird, William (1822), *A Geological Survey of the Yorkshire Coast*, Clark, Whitby

## chapter one:
## *Dangerous Rocks*

Young, George (1817), *A History of Whitby*, Clark & Medd, Whitby

## chapter two:
## *But Dreames*

I am grateful to Emilie Savage-Smith of the Wellcome Unit for the History of Medicine, Oxford for information and a string of references on this subject.

There is an unproven claim for an even earlier alum industry in this area. In his *History of Whitby* (1817), George Young mentions a field near Grosmont, about six miles inland from Whitby, known as Allum-Garth. The field contains the outlines of buildings which could have been alum houses. If so, these were either Roman or monastic works, far predating the seventeenth-century works on the coast. Young speculates that the Romans knew how to make alum from shale, but that the technique was lost during 'the iruption of the Goths'.

R. B. Turton (1938) makes an attempt, which is only partially successful, to unravel the intricate history of the Chaloner family. He concludes that the original founder of the alum works was not the same man who sat in judgment on Charles I. Even if this is so, the perceived wrong done to the Chaloner family remains a possible factor in their political allegiances.

For Chinese sources from pp.4 to 8, see Needham, Joseph.
Agricola, Georgius (1556) *De Re Metallica,* translated with annotations and appendices by H. C. and L. H. Hoover, published by The Mining Magazine, London 1912; facsimile reprint published by Dover, New York in 1950, Book XII, pp. 564–72
Anawati, Georges (1996) 'Arabic alchemy', in Roshdi Rashed (ed.) *The Encyclopedia of the History of Arabic Science,* vol. 3, Routledge, London
Aubrey, John (1813) *Lives of Eminent Men, published from original manuscripts in the Ashmolean Museum,* Oxford
Bacon, Francis (1626) *Naturall History,* William Lee, London

Kendall, Percy Fry and Wroot, Herbert E. (1924) *The Geology of Yorkshire*, 2 vols., P. F. Kendall (self-published), Leeds

Marshall, Gary (1995) 'Redressing the balance – an archaeological evaluation of North Yorkshire's coastal alum industry', *Industrial Archaeology Review*, vol. 18, no. 1, pp. 39–62

Needham, Joseph (1954 onwards) *Science and Civilisation in China*, vol. 5, part 3, Cambridge University Press, Cambridge

Rackham, H. (ed. and transl.) (1938–63) *Pliny, Natural History*, 10 vols, Heinemann, London, extract from Book 35

Rhazes, see Steele, Robert

Sabra, A. I. (1976) 'The scientific enterprise' in B. Lewis (ed.) *The World of Islam*, Thames and Hudson, London

Steele, Robert (1929) 'Practical chemistry in the twelfth century: *Rasis de alumnibus et salibus*, translated [into Latin] by Gerard of Cremona.' *Isis*, vol. 12, pp. 10–46. English translation by Dilys Cluer, 1997.

Turton, R. B. (1938) *The Alum Farm*, Horne & Son, Whitby. Reissued 1987 by Michael Moon, Whitehaven

Young, George (1817) *A History of Whitby*, Clark & Medd, Whitby

## chapter three:
## *The First Reptile*

British Museum (1910) *A Guide to the Fossil Reptiles, Fishes and Amphibians in the Dept of Geology and Palaeontology in the British Museum (Natural History)*, 9th edn, British Museum, London

Chapman, William (1758) 'An account of the fossile bones of an allegator, found on the sea-shore near Whitby in Yorkshire', *Philosophical Transactions of the Royal Society*, vol. 50, pp. 688–691

Wooler (1758) 'A description of the fossil skeleton of an animal found in the Alum Rock near Whitby', *Philosophical Transactions of the Royal Society*, vol. 50 pp. 786–790

## chapter four:
## *A Journey to a Birth*

This fictional piece is based on William Smith's own account of this journey as transcribed by his nephew John Phillips. Smith was accompanied on his tour by Samborn Palmer and Dr Perkins. The opening of the piece borrows its tone from T. S. Eliot's poem, 'Journey of the Magi'.

Phillips, John (1844) *Memoirs of William Smith*, John Murray, London
*York Directory* (1798), taken from *The Universal British Directory of Trade and Commerce, Volume IV*, Champante and Whitrow, London

## chapter five:
## *The Second Reptile*

Young, George (1820) 'Account of a singular fossil skeleton; discovered at Whitby, in February 1819', *Memoirs of the Wernerian Natural History Society*, vol. 3 pp. 450–457
Young, George (1838) *Scriptural Geology*, 2 vols, Simpkin Marshall, London

## chapter six:
## *Stones from the Sky*

Some readers might wonder why stories of meteorites are included in a book about the history of geology. The answer is that geology is the study of the Earth, both as a whole and of its constituent parts. Geology starts with the formation of the Earth and solar system and meteorites are an essential clue to how this came about. Meteorites are therefore 'claimed' by geologists, as well as by planetary and space scientists.

*Preamble*

Santillana, Giorgio de (1961) *The Origins of Scientific Thought*, University of Chicago Press, Chicago

*Stories from the Field of Heaven*

This chapter draws on the 1994 paper by Ursula Marvin of the Harvard-Smithsonian Center for Astrophysics. Dr Marvin in turn fully acknowledged her own debt to the work of A. Alvarez published in 1926. I am grateful to Dr Marvin for supplying a copy of her paper and for her helpful suggestions. Any errors are of course my own responsibility.

The Campo del Cielo seems to be an endless source of fascinating stories. Here are three brief tales which could not be fitted into the body of the book.

In 1816 a gunsmith in Buenos Aires was asked to make a pair of flintlock pistols from a piece of a meteorite from the Campo del Cielo. These exquisitely ornamented pieces were presented to James Madison, fourth President of the United States. Wishing to avoid diplomatic involvement in Argentina's conflict with Spain, Madison passed the gift on to his Secretary of State, James Monroe. The pistols eventually found their way to the Monroe Museum in Fredericksburg, Virginia. Thirty years ago minute samples taken from the pistols showed that their metal contained no nickel, and therefore could not have come from the meteorite. Ever since, it has been generously assumed that the gunsmith found the meteorite metal to difficult to work, and substituted wrought iron. It seems he saw fit to keep this to himself; his secret remained intact for 150 years.

In 1826 the British Foreign Secretary, George Canning, sent Sir Woodbine Parish as an emissary to the government of the new state of Argentina. In reply to this friendly gesture the Argentine government presented Sir Woodbine with an extraordinary gift. Not a priceless piece of jewellery, nor an exotic animal, but a massive piece of a meteorite. Sir Woodbine was grateful, but the gift brought its own difficulties. The piece weighed over 600 kilograms (1,300 pounds), and closely resembled a mass of solid iron. Sailors immediately suspected that it must be a lodestone, a magnet of enormous power. Such a stone would, they believed, draw the nails out of their ships and disturb their compasses. They would not take it on board. Though the piece was put into a wooden box, its weight was so extraordinary that there was no way of disguising it. No merchant crew would take it, so a man of war of the Royal Navy was dispatched, and the huge iron mass made its 12,000-mile journey to London.

The largest meteorite from the Campo del Cielo, the Chaco iron, was discovered in the 1960s using a metal detctor. It was excavated from the ground by

Argentine soldiers and left on the surface of its crater. An American collector and dealer in geological specimens, Robert Haag, arranged the purchase of the meteorite from an Argentinian mineralogist. In 1990 he arrived at the site with a portable derrick, hoisted the 33-ton lump of iron on to a flatbed truck, and headed for home. He was arrested at the provincial border, and thrown in jail for the offence of illegally attempting to take a meteorite out of the country. He posted $20,000 bail and left the country. As far as we know his case is still pending – though the meteorite is now back where it belongs.

Alvarez, A. (1926) *El Meteorito del Chaco*, Casa Jacobo Peuser, Buenos Aires

Cassidy, William et al. (1965) 'Meteorites and craters of Campo del Cielo, Argentina', *Science*, vol. 149, no. 3688

Marvin, Ursula B. (1994) 'The meteorite of Campo del Cielo, Argentina: its history in politics, diplomacy and science', pp. 155–174 in *Geological Sciences in Latin America, Scientific Relations and Exchanges*, S. Figueiroa and M. Lopes (eds) Univ. Estad Campinas, São Paolo, Brazil

Rubin de Celis, Don Michael (1787) 'An account of a mass of native iron found in South America', *Philosophical Transactions of the Royal Society*, vol. 78, pp. 37–42 (Spanish); pp. 183–189 (English translation)

*An obsession with truth*

The theme of this chapter, though not the style, was inspired by the painstaking research into the life of Edward Topham carried out by C. T. and J. M. Pillinger, as detailed in their 1996 paper – though in each case I have gone back to the original source material. For anyone wishing to know more, Professor Pillinger is currently completing a book on the Wold Cottage meteorite. I am grateful to Dr Robert Hutchinson of the Natural History Museum for pointing me in the right direction. His name, in turn, came to light in a *Guardian* article by Martin Wainwright marking the 200th anniversary of the fall of the Wold Cottage meteorite.

Sir Humphry Davy's reference to Piazzi's work (p. 94) is a reminder of Bode's Law, which showed that the distance of each planet from the Sun followed a simple mathematical series. There was though a 'missing' planet between Mars and Jupiter, which astronomers set out to find. Piazzi and then Olbers found 'minor planets' or asteroids in the right place, and the law was confirmed.

The discovery of Neptune broke the rule, however, and Bode's Law was quietly forgotten.

Banks, Sir Joseph (1794) Letter to Sir William Hamilton, enclosing letter from the Earl of Bristol, Bishop of Derry, Egerton collection, British Museum manuscript collection. MS 2641, 153–154

Burke, J. G. (1986) *Cosmic Debris: Meteorites in History*, University of California Press, Berkeley

Bynum, W. F., Browne, E. J. and Porter, R. (eds) (1981) *Dictionary of the History of Science*, Macmillan, London

*Gentleman's Magazine* (1768) vol. 38, p. 539

*Gentleman's Magazine* (1783) vol. 53, pp. 443 & 485

*Gentleman's Magazine* (1797) vol. 67, pp. 549–551

*Guardian*, (1995) 14 December, 'Scientists honour "stone from sky"'

Hamilton, Sir William (1795) 'An account of the eruption of Mt Vesuvius', *Philosophical Transactions of the Royal Society*, vol. 85, pp. 73–116

Howard, E. C. (1802) 'Experiments and observations on certain stony and metalline substances which at different times are said to have fallen on the earth', *Philosophical Transactions of the Royal Society*, vol. 92, pp. 168–212

King, Edward (1796) *Remarks concerning Stones said to have Fallen from the Clouds both in These Days and in Antient Times*, G. Nichol, London

*London Chronicle* (1796) 7–9 January, Anonymous letter from Sheffield dated 1 January

Pillinger, C. T. and Pillinger, J. M. (1996) 'The Wold Cottage meteorite: not just any ordinary chondrite', *Meteoritics & Planetary Science*, vol. 31, pp. 589–605

*Public Characters* (1804) vol. 7, pp. 198–212

Reynolds, Frederic (1826) *The Life & Times of Frederic Reynolds*, Colburn, London

Sears, D. W. (1975) 'Sketches in the history of meteoritics 1: the birth of the science', *Meteoritics*, vol. 10, no. 3, pp. 215–225

Sumbel, L. (1811) *Memoirs of the Life of Mrs Sumbel (late Wells)*, Chapple, London

Topham, Edward (1776) *Letters from Edinburgh*, J. Dodesley, London

Weld, Charles (1848) *A History of the Royal Society*, Royal Society, London, 2 vols

## chapter seven:
## *The Third Reptile*

My thanks to Rosemary Roden and to Kate Andrew for information and advice during their time at Whitby Museum.

Andrew, K. (1997) Public notice displayed in Whitby Museum during cleaning of the specimen.
Browne, H. B. (1946) *Chapters of Whitby History*, Brown & Sons, Whitby
Whitby Literary and Philosophical Society (1825), 2nd annual report
Whitby Literary and Philosophical Society (1827), 4th annual report
Young, George (1825) 'Account of a fossil crocodile recently discovered in the alum-shale near Whitby', *Edinburgh Philosophical Journal*, vol. 13, p. 76

## chapter eight:
## *Whitby to the World*

I am extremely grateful to Walter Robinson of Hanover Arts Bookshop in Scarborough for digging out various books of ancient maps from his shelves, and for first alerting me to the map used on the cover of the book. John Phillips of Scarborough kindly supplied the print.

Beaglehole, J. C. (ed.) (1955) *The Journals of Captain James Cook on his Voyages of Discovery, Vol 1 The Voyage of the Endeavour, 1768–1771*, Cambridge University Press and the Hakluyt Society, Cambridge
Beaglehole, J. C. (1974) *The Life of Captain James Cook*, A. & C. Black, London
Calasso, Roberto (1988, trans. 1993) *The Marriage of Cadmus and Harmony (Le nozze di Cadmo e Armonia)*, Jonathan Cape, London
Conrad, Joseph (1902) *Heart of Darkness*, reprinted by Wordsworth Classics, London
Cook, James, see Beaglehole (1955)
Homer (1946) *The Odyssey*, trans. E. V. Rieu, Penguin Books, Harmondsworth
Ptolemy, Claudius (*c*.140) *Cosmographia, Book VII*
Santillana, Giorgio de (1961) *The Origins of Scientific Thought*, University of Chicago Press, Chicago

Schaffer, Simon (1996) *London Review of Books*, 31.10.96
Shapin, Steven (1996) *The Scientific Revolution*, University of Chicago Press, Chicago
*Whitby Gazette, Captain Cook's Endeavour, the Whitby Homecoming* (1997), special publication, Whitby Gazette, Whitby

## chapter nine: *Devils, Romans and Rocks*

Brumhead, Derek (1979) *Geology Explained in the Yorkshire Dales and on the Yorkshire Coast*, David & Charles, Newton Abbot
Kendall, Percy Fry and Wroot, Herbert E. (1924) *The Geology of Yorkshire*, 2 vols., P. F. Kendall (self-published), Leeds
Parkinson, Thomas (1888) *Yorkshire Legend and Tradition*, Elliot Stock, London
Simpson, Martin (1855) *The Fossils of the Yorkshire Lias, Described from Nature*, Whittaker, London

## chapter ten: *The Fourth Reptile*

Dunn, John (1831) 'On a large species of Plesiosaurus in the Scarborough Museum', *Proceedings of the Geological Society of London*, vol. 1, pp. 336–337
Scarborough Philosophical Society (1831) Report of the Council, 31 August
Williamson, W. C. (1837) 'On the distribution of fossil remains on the Yorkshire coast', *Transactions of the Geological Society of London*, series 2, vol. 5, pp. 223–242
Young, George and Bird, John (1828) *A Geological Survey of the Yorkshire Coast*, 2nd edn, Kirby, Whitby

chapter eleven:

## *Geological Journeys*

Buckland, William (1836) *Geology and Mineralogy Considered with Reference to Natural Theology (Bridgewater Treatise, no. 6)*, William Pickering, London

Conybeare, William and Phillips, William (1822) *Outline of the Geology of England and Wales*, William Phillips, London

Fortey, Richard (1993) *The Hidden Landscape*, Jonathan Cape, London

Phillips, John (1829) *Illustrations of the Geology of Yorkshire*, Wilson, York

chapter twelve:

## *Fossils, Debts and Friends in High Places*

This chapter is based on a marvellous and meticulous piece of research by the scientific historian, the late Joan Eyles. I came across her paper on the sale of William Smith's fossil collection in the local studies section of Scarborough Public Library. I am indebted to Joan Eyles and to the anonymous Scarborough librarian(s) who had the foresight to acquire her paper, together with an impressive collection of books on the geology of the area. Any extracts from documents not taken from Joan Eyles's paper are from William Smith's manuscript diaries, in the University Museum, Oxford.

Eyles, Joan M. (1967), 'William Smith: the sale of his geological collection to the British Museum', *Annals of Science*, vol. 23, no. 3, pp. 177–212

chapter thirteen:

## *The Fifth Reptile*

This story, and Adam Sedgwick's correspondence, are referred to in Benton and Taylor's 1984 paper, which was the starting point for my interest in the Cambridge plesiosaur. I would like to thank the staff of the Manuscript Room of Cambridge University Library for their courteous and efficient help.

I am also grateful to Mike Dorling of the Sedgwick Museum, Cambridge who took time to tell me the recent history of the specimen during my visit to Cambridge, and supplied copies of the Whitby poster and a portrait of Adam Sedgwick.

Benton, M. J. and Taylor, M. A. (1984) 'Marine reptiles from the Upper Lias (Lower Toarcian, Lower Jurassic) of the Yorkshire coast', *Proceedings of the Yorkshire Geological Society*, vol. 44, part 4, pp. 399–429

Clark, J. W. and Hughes, T. McK. (1890) *The Life and Letters of the Reverend Adam Sedgwick*, vol. 2, pp. 37–38, Cambridge University Press, Cambridge

*Reed's Illustrated Guide To Whitby* (1854), Reed, Whitby

Sedgwick, Adam: Manuscript correspondence, Cambridge University Library Add. MSS 7652/ID,IE and 7652/ID/111d (By kind permission of the Syndics of Cambridge University Library)

Seeley, H. G. (1865) 'On *Plesiosaurus macropterus*, a new species from the Lias of Whitby', *Annual Magazine of Natural History*, ser. 3, vol. 15, pp. 49–53 & 232–233

## chapter fourteen:
## *Geology and Repose*

Simpson, Martin (1855) *The Fossils of the Yorkshire Lias, Described from Nature*, Whittaker, London

## chapter fifteen:
## *The Sixth Reptile*

I am grateful to the present Marquis of Normanby, to Miss F. J. E. Moorhead, Assistant Archivist, Mulgrave Castle Archives, to Barbara Pyrah and to Sally Kingston of the Yorkshire Philosophical Society, for their generous help, and for the information they provided about the history of this fossil. There is some interesting historical confusion over the name of the Yorkshire Philosophical Society and the Yorkshire Museum. When these were first set up in York, other towns in Yorkshire objected to the adoption of the county's name for a town society and museum. Because of this understandable chauvinism, and because the society and museum are based in York, they are often referred to in documents as the York Philosophical Society and York Museum.

Charlesworth, Edward (1845) 'Notice of the discovery of a large specimen of *Plesiosaurus* found at Kettleness, on the Yorkshire Coast', *Annual Report of the British Association, 1844*

Normanby, Marquis of (1845) Manuscript correspondence, T/276, Mulgrave Castle, Whitby

Pyrah, Barbara (1988) *The History of the Yorkshire Museum and its Geological Collections,* William Sessions, York

Yorkshire Philosophical Society (1845) Council Minute Book, 8 March 1845 and 31 March 1845

## chapter sixteen:

## *The Strange Case of the Hyenas' Bones*

My sincere thanks to Paul Ensom, Curator of Geology at the Yorkshire Museum, for taking the time to show me the museum's collection of bones from the Kirkdale cave, and for an interesting discussion on both their present condition and William Buckland's own description of them. The secondary sources were extremely valuable in giving background to this story (in particular Rupke's account of the in-fighting at Oxford). However, the most persuasive account of the Kirkdale cave is Buckland's own, as given in his book *Reliquiae Diluvianae,* a brilliant example of science as argument for ideas. Many of the passages used in the summary of the case (where Foster is reading Buckland's words) are taken directly from Buckland's book. The format of this piece follows that of the classic Sherlock Holmes story. Some specifics, for example Foster's previous army career, and the first meeting of the two principals in a laboratory, are an echo of the first Holmes story, *A Study in Scarlet.* Other minor details recur in almost all the Holmes stories, for example Holmes's irritation at the destruction of evidence at the scene of the crime, the proposal of an obvious and incorrect solution by an inferior investigator, and Holmes's peculiar hound-like behaviour while searching for clues.

Boylan, Patrick (1967) 'Dean William Buckland, 1784–1856: a pioneer in cave science', *Studies in Speleology*, vol. 1, pp. 236–253
Boylan, Patrick (1972) 'The scientific significance of the Kirkdale cave hyenas', *Yorkshire Philosophical Society Annual Report for 1971*, pp. 38–47
Boylan, Patrick (1981) 'A new revision of the Pleistocene mammalian fauna of Kirkdale cave, Yorkshire', *Proceedings of the Yorkshire Geological Society*, vol. 43, part 3, no. 14, pp 253–278
Buckland, William (1823) *Reliquiae Diluvianae*, John Murray, London
Conan Doyle, Arthur (1887) *A Study In Scarlet*, Lippincott, London
Rupke, Nicholas (1983) *The Great Chain of History*, Clarendon, Oxford
Santillana, Giorgio de (1961) *The Origins of Scientific Thought*, University of Chicago Press, Chicago

## chapter seventeen:
## *The Seventh Reptile*

Nigel Monaghan of the National Museum of Ireland provided crucial assistance in the research of this piece, for which I am extremely grateful.

Carte, Alexander and Baily, W. H. (1863) 'Description of a new species of Plesiosaurus, from the Lias, near Whitby, Yorkshire' *Journal of the Royal Dublin Society*, vol. 4, pp. 160–170
Monaghan, N (1992) 'Geology in the National Museum of Ireland', *Geological Curator*, vol. 5, no. 7, pp. 275–282
Monaghan, N (1997) Letter to author, dated 14 March
Phillips, John (1853) 'Report of the Council of the Yorkshire Philosophical Society', in *Annual Reports of the Yorkshire Philosophical Society, 1852*
Zoological Society of Ireland (1853) 'Report of the Council', in *Annual Reports of the Zoological Society of Ireland*
Zoological Society of Ireland (1859) 'Report of the Council', *Annual Reports of the Zoological Society of Ireland*

chapter eighteen:

## *Embracing and Uniting*

Simpson, Martin (1855) *The Fossils of the Yorkshire Lias, Described from Nature,* Whittaker, London

chapter nineteen:

## *The Eighth Reptile*

Browne, H. B. (1946) *Chapters of Whitby History, 1823–1946,* Brown & Sons, Whitby

chapter twenty:

## *Thinking in Four Dimensions*

Alexander, Jan (1986) 'Idealised flow models to predict alluvial sandstone body distribution in the Middle Jurassic Yorkshire Basin', *Marine and Petroleum Geology,* vol. 3, November, pp. 298–305

Bird, Charles (1881) *A Short Sketch of the Geology of Yorkshire,* Simpkin, Marshall & Co., London

Brumhead, Derek (1979) *Geology Explained in the Yorkshire Dales and on the Yorkshire Coast,* David & Charles, Newton Abbot

Fox-Strangways, C. and Barrow, G. (1915) 'Geology of the country between Scarborough and Whitby', *Memoir of the Geological Survey of Great Britain*

Harris, T. M. (1953) 'The Geology of the Yorkshire Flora', *Proceedings of the Yorkshire Geological Society,* vol. 29, pp. 63–71

Harris, T. M. (1942–53) 'Notes on Yorkshire Jurassic flora', *Annals and Magazine of Natural History,* vol. 11, pp. 9–13; vol. 12, pp. 1–6

Hemingway, J. E. (1953) 'Report of a field meeting at Whitby held in June 1949', *Proceedings of the Yorkshire Geological Society,* vol. 28, pp. 118–122

Hemingway, J. E. (1958) 'The geology of the Whitby area' in *A Survey of Whitby and the Surrounding Area*, G. H. J. Daysh (ed.), Shakespeare Head, Eton, Windsor

Hemingway, J. E. and Wilson, V. (1963) *Geologists Association Guide, No. 34: The Yorkshire Coast*, Benham, Colchester

Hemingway, J. E. (1974) 'Jurassic' in *The Geology and Mineral Resources of Yorkshire*, Yorkshire Geological Society, Leeds

*Horne's Guide to Whitby* (1904), 9th edn, Horne, Whitby

Kantorowicz, J. D. (1990) 'Early diagenesis, Ravenscar Group', *Proceedings of the Yorkshire Geological Society*, vol. 48, part 1, pp. 61–74

Kendall, Percy Fry and Wroot, Herbert E. (1924) *The Geology of Yorkshire*, 2 vols., P. F. Kendall (self-published), Leeds

Phillips, John (1829) *Illustrations of the Geology of Yorkshire*, Wilson, York

Rastall, R. H. and Hemingway, J. E. (1940) 'The Yorkshire Dogger', *Geological Magazine*, vol. 77, pp. 177–197

Walmsley, Leo (1919) *Fossils of the Whitby District*, Horne, Whitby

Young, George (1817) *A History of Whitby*, Clark & Medd, Whitby

Young, George and Bird, William (1822), *A Geological Survey of the Yorkshire Coast*, Clark, Whitby

## chapter twenty-one:

## *The Ninth Reptile*

Collier, Frederick J. (1997) Letter to author, dated 25 March

Cunningham, C. (1997) Letter to author, dated 21 June

Houston Museum of Natural Sciences (1961) *Museum News*, December

Marshall, John P. (1876) Letter to Board of Trustees of Tufts College, 29 April 1876, University Archives, Tufts University

Scheele, William (1961) Letter to T. E. Pulley, dated 19 June, Houston Museum of Natural Sciences chapter twenty-two:

## *Yorkshire Ammonites*

### *Saints and geologists*

Charlton, Lionel (1779) *The History of Whitby*, A. Ward, York
Pevsner, Nikolaus (1966) *The Buildings of England: Yorkshire, the North Riding*, Penguin, London
Phillips, John (1829) *Illustrations of the Geology of Yorkshire*, Wilson, York
Simpson, Martin (1855) *The Fossils of the Yorkshire Lias, Described from Nature*, Whittaker, London
Smith, William (1836) Personal papers, Box 46 Folder 1, Oxford University Museum, Oxford

### *A life's work*

I am grateful to Christopher Toland for first telling me about Louis Hunton and his work, and for providing these references.

Hunton, Louis (1837) 'Remarks on a section of the Upper Lias and marlstone of Yorkshire, showing the limited vertical range of the species of ammonites, and other testacea, with their value as geological tests', *Transactions of the Geological Society of London*, series 2, vol. 5, pp. 215–221
Torrens, H. S. and Getty, T. A. (1984) 'Louis Hunton (1814–1838) – English pioneer in ammonite biostratigraphy', *Earth Sciences History*, vol. 3, no. 1, pp 58–68

chapter twenty-three:

## *The Tenth Reptile*

Broadhurst, Frederick M. and Duffy, Louis (1970) 'A plesiosaur in the Geology Department, Manchester University', *Museums Journal*, vol. 70, pp. 30–31

## chapter twenty-four:

## *A Museum at Scarborough*

I am grateful to Stella Brecknell of the University Museum, Oxford for the chance to look through William Smith's manuscript papers, and to Professor Hugh Torrens of the University of Keele and Sophie Forgan of Teesside University for advice and information. My thanks also to Dr Jane Mee, head of Scarborough Museums Service and to Paul Parry for allowing me access to the museum archives.

Scarborough Philosophical Society (1830) *First Annual Report for 1829*, Todd, Scarborough

Sharpe, R. H. (1829) 'Architect's drawings for a museum at Scarboro'. Unpublished document, Woodend Museum of Natural History, Scarborough

Smith, William (1804–5?) Personal papers, Box 13 Folder 4, Box 41 Folders 1 to 5, Oxford University Museum, Oxford

## chapter twenty-five:

## *Ice, Water and Landscape*

Kendall, Percy Fry (1902) 'A system of glacier-lakes in the Cleveland Hills', *Quarterly Journal of the Geological Society of London*, vol. 58, pp. 471-571

Kendall, Percy Fry and Wroot, Herbert E. (1924) *The Geology of Yorkshire*, 2 vols., P. F. Kendall (self-published), Leeds

# Copyright acknowledgements

## ILLUSTRATIONS

p. 3 Photograph by F. M. Sutcliffe, copyright Sutcliffe Gallery, Whitby, in association with, and by kind permission of, Whitby Literary and Philosophical Society.

p. 19 Photograph by T. Watson, copyright Whitby Literary and Philosophical Society, reproduced with kind permission.

p. 63 Portrait of Edward Topham by John Russell, courtesy Christie's Images.

p. 114 Artist unknown, reproduced by kind permission of Whitby Literary and Philosophical Society.

p. 136 Photograph by John Tindall, copyright John Tindall, Whitby.

p. 153 Portrait of William Smith by H. Forau, reproduced by kind permission of the Geological Society.

p. 180 Reproduced by kind permission of the Sedgwick Museum, Cambridge.

p. 190 Portrait of Adam Sedgwick, 1833 mezzotint, reproduced by kind permission of the Geological Society and the Sedgwick Museum, Cambridge.

p. 220 Portrait of William Buckland by R. Ansdell, *c.*1843, reproduced by kind permission of the Geological Society.

p. 267 Photographer unknown, reproduced by kind permission of Whitby Literary and Philosophical Society.

p. 295 Photograph by F. M. Sutcliffe, copyright Sutcliffe Gallery, Whitby, in association with, and by kind permission of, Whitby Literary and Philosophical Society.

pp. 315–17 Drawings of the Rotunda Museum by R. H. Sharp, reproduced by kind permission of Scarborough Museums & Gallery, Department of Tourism and Leisure Services, Scarborough Borough Council.

## TEXT EXTRACTS

pp. 4–8 By kind permission of Cambridge University Press.
p. 140 From *Geology Explained in the Yorkshire Dales and on the Yorkshire Coast* (David & Charles, 1979) by kind permission of the publishers.
pp. 206–8 By kind permission of the Marquis of Normanby.
p. 283 From *Geology Explained in the Yorkshire Dales and on the Yorkshire Coast* (David & Charles, 1979) by kind permission of the publishers.

All other illustrations and text extracts have been fairly assumed to be out of copyright, or are acknowledged elsewhere or are the work of the author. If there are any exceptions, or any inadvertent breaches of copyright, please accept our apologies and contact the publisher.

# Index